Do-It-Yourself Plumbing...

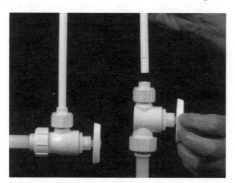

It's easy with Genova

Do-It-Yourself Plumbing
It's easy with Genova

Copyright© 1987, 1991, 2000
All rights reserved
Printed in the United States of America

Edited by:
Robert M. Williams

Photography by:
Richard Day

Produced by:
Genova Products, Inc.
7034 East Court Street
Davison, MI 48423

Library of Congress Cataloging-in-Publications Data

Day, Richard, date
  Do-It-Yourself Plumbing

  1. Plumbing--Amateurs' manuals. I. Genova Inc.
II. Title
TH6124d356   1986   696'.1   86-14871
ISBN 0-9616509-0-7

Genova Products, Inc.
7034 East Court Street
Box 309
Davison, MI 48423-0309

Form No. 253326

Helpful Hints from Genova

Plumbing vs. Electricity

The home plumbing and electrical systems serve totally different purposes. Yet they have something in common, something that every home plumber should know: **plumbing and electricity make a potentially dangerous combination.**

Very little around the home is better grounded than a metal plumbing system. And when touching any of its metal parts, you are well grounded, too. If, at the same time, you are working with a faulty electrical tool or appliance, an electrical current may flow through your body to ground. The consequences could be disastrous.

The good news is that due to Genova's all-thermoplastic plumbing system, the danger of becoming grounded by the plumbing is very greatly reduced. This is because plastic piping is a nonconductor of electricity. Contact with a thermoplastic pipe or fitting does not ground you. However, even in an all-plastic plumbing system, metal plumbing fixtures and appliances—such as sinks—may be grounded by something other than the piping. For example, a kitchen sink with a food waste disposer is more than likely grounded through the disposer's electrical wiring.

So, to be safe when working with electrical tools and equipment, keep away from **all** metal plumbing parts. For the same reason, it is recommended that a flashlight, instead of a drop light, be used around a metal plumbing system.

Another danger of plumbing vs. electrical systems is water. Water and electricity don't mix. If the skin is wet when contacting a live conductor, much more current will flow than if the skin is dry. This makes it dangerous to work with power tools when damp or wet, when standing in water, or when your hands are sweaty. Work dry.

Not Just for Plumbing Any More

Today you will find many uses for Genova vinyl pipe and fittings beyond plumbing. With a little pipe and a lot of creativity the uses are endless. One idea is to construct your own recycling center, using PVC pipe to separate newspapers, glass, and plastics. Pipe furniture is an old favorite for decks and porches because of its stability and resistance to rust, but it can also be used in the home as inexpensive bookcases, plant stands, and, of course, couches and chairs. Imagine the furniture on a small scale for dolls. Other creative ideas include frames for practice nets or tarps, and garbage trolleys. Vinyl pipe can be used in hammocks, small greenhouses, curtain valances, playhouses, coat racks, etc. -- and these are only a portion of what we've seen made with PVC. For more information, simply call Genova or talk to your do-it-yourself retailer.

Because it's sturdy, waterproof, and easy to assemble, Genova vinyl pipe is the obvious choice for the inventive mind.

Beyond the House

Genova products also find many uses on boats, travel trailers, motor and mobile homes. Possible projects might include a dockside water system, a pressurized RV water supply system, adding a hot water heater system or shower, fixture modernization, conversion to self-contained, or complete water supply plumbing using vinyl tubing. Vinyl is ideal for marine/RV water supply. Easy to use, it can be run through bulkheads almost like electrical wiring, and because the tube is resistant to freeze damage, vinyl offers unsurpassed reliability. What's more, you'll find Genova's Universal Fittings a handy way to make quick leak-free connections.

Geno Hotline

As a service to readers of this book who become do-it-yourself home plumbers, Genova offers the toll-free **Geno Hotline** Club. As a member of this group, if you have a question on plumbing, simply dial 1-800-521-7488. Ask for "Geno The Plumber." Your call will be routed to a knowledgeable "Geno," who is backed up by Master Plumber Robert M. Williams, chairman and chief executive officer of Genova. In most cases, "Geno" can answer your plumbing questions on the spot. But if it takes research to give you an authoritative reply, that will be done. Then "Geno" will call you back as soon as he can. Your question may be on any phase of home plumbing, whether with thermoplastic piping or some other kind. The call is free to readers of this book, and so is the **Geno Hotline** service. It's part of Genova's motto: "The People Who Get it Done." (Of course, "Geno"—like Santa Claus—has helpers. But you get the idea.)

Getting Genova Products

To get Genova products, in most cases, you can walk into your local dealer and pick out what you need from the display racks. If a Genova product you need isn't stocked by your dealer, it can be ordered for you. Most major home centers, lumber yards, chain stores, and hardware stores handle Genova products. Make sure to ask for them by name.

Or feel free to call the **Geno Hotline** number to ask for Genova's free mail order catalog.

Genova's vow is that anything you need will be made available to you.

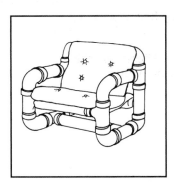

Table of Contents

Section 1 House Plumbing Spelled Out
Chapter 1 The Home Plumbing System

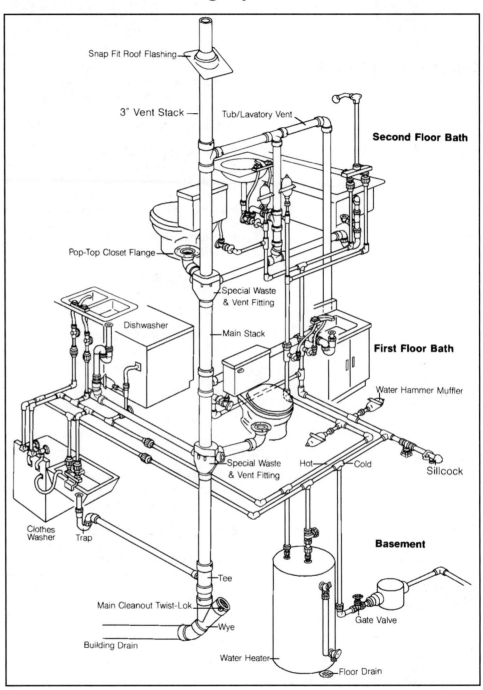

Snap Fit Roof Flashing

3" Vent Stack

Tub/Lavatory Vent

Second Floor Bath

Pop-Top Closet Flange

Special Waste & Vent Fitting

Dishwasher

Main Stack

First Floor Bath

Water Hammer Muffler

Special Waste & Vent Fitting

Hot — Cold

Sillcock

Clothes Washer   Trap

Basement

Tee

Main Cleanout Twist-Lok

Wye

Gate Valve

Building Drain

Water Heater

Floor Drain

1-1 The complete home plumbing system

Your home plumbing system is two separate systems: water supply and water disposal. The water disposal system is called the Drain-Waste-Vent (DWV) system.

Both systems are made up of several hundred feet of pipes with many fittings to join them. These bring the water into the house and distribute it to every plumbing appliance and fixture. They then collect the wastes from each, carrying them out of the house to the sewer or septic system (Fig. 1-1). Plumbing systems are

Genova's business. The firm makes more than 700 different types of pipe and fittings to plumb a whole house, as well as the sewer and septic system.

The water supply and DWV systems are never connected to each other. Such cross connections, as they're called, are outlawed by all plumbing codes. This is because contamination getting into drinking water through a cross connection can sicken or even kill.

1-2a

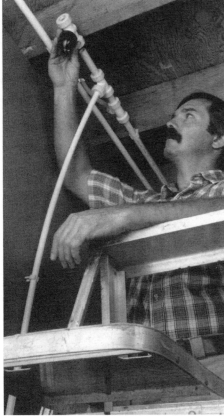

1-2b

1-3

1-2a, b Two ways to plumb: When you use easy-working thermoplastic pipes and fittings that join easily, home plumbing becomes one of the simplest home projects you can do. Genova makes all the rigid and flexible pipes and fittings to help you get it done. And you can choose either Genova's solvent welding system (1-2a) or Genova's mechanically coupled Universal Fittings (1-2b).

1-3 "Plumbing" gets its name from the word *plumb*, meaning, vertical or straight up and down. Also, *plumb* is part of the Latin word for "lead" which was the primary metal first used in plumbing pipes. When doing plumbing projects, a level is used to get the desired piping alignments.

DIY Plumbing

When the right materials are used--chiefly PVC vinyl--plumbing is simple. You **can** do it yourself. Lots of people do. In fact, if you can wield a paint brush, swing a saw, if you know how to read a rule, use a level, and you know water flows downhill, you will be a big success at do-it-yourself plumbing.

By making the installation yourself, you stand to save about three-fourths of the cost. This is because plumbing is about 25 percent materials and 75 percent labor. The exact 25:75 mix varies slightly according to the level--the quality--of fixtures used. Another benefit of do-it-yourself plumbing: if you run out of money, you can stop and continue later.

1-4

1-5

1-4 Control of pressurized water in the home distribution system relies on strategically placed valves. These universal line valves have side waste openings that are plugged with screws. When the hot/cold water valve handles are closed, the pipes may be drained by removing the screws.

1-5 Here are the tools needed for home plumbing remodeling and additions, or for plumbing a whole house with easy-to-handle thermoplastic piping. They are: level, 3/8″ drill, pipe wrench, measuring tape, drill bits, hammer, chisel, fixture wrench, fine-tooth saw, trap wrench, eye goggles, knife, and an adjustable open-end wrench.

Water supply system

There are two kinds of plumbing: rough and finish. Rough plumbing is done before the walls are closed in. This involves installing the pipes and fittings. Finish plumbing is done after the walls have been completed. It involves installation of the plumbing fixtures and appliances, plus connecting them to their piping. Planning for both kinds should be done up front.

Each adult in a family and child over two years old uses from 50 to 100 gallons of water a day. Your house water supply system either gets this water from the city main, which is usually buried near the street, or from a domestic well. The house water supply system delivers water around the house to the various fixtures, appliances, and water outlets. The piping necessary to do this varies in size from 3/4″ or 1″ in diameter down to only 1/4″. Where much water must flow, the pipes must be larger. Where little water must flow, the pipes can be

smaller. The average home water supply system is pressurized to around 50 psi. This is why water supply pipes can be small compared to drain-waste-vent pipes, yet still carry enough water.

Main shutoff valve. After entering your property, water goes through a below-ground utility shutoff valve, then an anti-backflow device (newer systems), then a water meter. Next, it comes to a valve called the house main shutoff. In most frost-belt homes, the main shutoff is located inside the house. It lets you turn off water to the whole house. The main shutoff may be located at the lowest point of the water supply system and contain a drain opening. Such a valve is called a universal line valve. With it, you can drain the whole system.

The valve of choice is the 3/4″ gate valve (Part No. 530271). It is the only valve available that gives full flow in a 3/4″ thermoplastic hot/cold water supply system and is compatible with CPVC, and copper tubing. What's more, the valve serves as a handy take-apart union. This feature may also be used in draining the plumbing system, giving this 3/4″ gate valve stop-and-waste capability. This Genova valve is pictured in Figs. 13-27, 28.

Hot/cold mains. Then, water runs through piping called the house cold water main. The first stop may be a water softener. Water for outdoor use bypasses the water softener via a hard water main. Branches offshoot to outside sillcocks.

The house cold water main leads to the water heater. This has a full flow valve where it enters the heater. The pipe leaving the water heater is the beginning of the hot water main. It runs parallel to the cold water main from that point on.

Branches of piping lead to groups of fixtures and to large water using

fixtures such as kitchen sinks, laundries, bathtubs, and showers. Pipes, mains or branches, going up vertically through walls are called risers. The water supply system ends at the fixtures and water-using appliances.

Changes from one kind of pipe to another are made with fittings called adapters, or transition fittings. For example, to go from a threaded steel pipe to a solvent welded one takes a threads-to-solvent-weld adapter. Differences in pipe size are accommodated by fittings called reducers. Some fittings combine a change of direction or a change in size along with some other function, such as in an angle adapter or a reducing tee.

Fixtures, appliances

Sinks, washbasins, toilets, and dishwashers should have shutoff valves on all water supplies. These are located on the wall or floor directly beneath the fixture. Known as supply valves (Fig. 1-7), they let you turn off water to the fixture in an emergency or for making replacements or repairs. The best ones also adapt the water supply system to reach the fixture's supply inlets. Sometimes adapters without valves are used, but it's better to have the valves, too. Either way, 3/8″ O.D. flexible tubes called riser tubes allow easy hookup between the water supply system and the fixture. Riser tubes also ease the changing of fixtures.

Water saving. In the interest of saving water, houses with modern plumbing contain water-saving faucets and shower heads. Conventional shower heads feed 6 to 8 gallons of water a minute. A water-saving shower head limits the flow to 3 gallons a minute, still giving a drenching, satisfying shower. It is estimated that replacing a

1-6

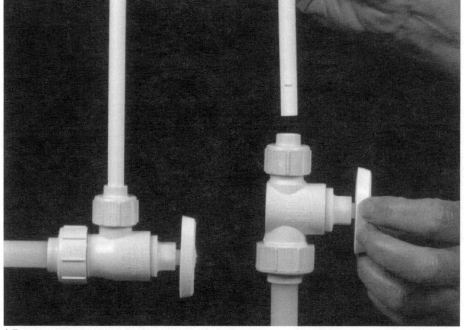

1-7

1-6 Flexible tubing not only makes neat, soldierly straight runs of tubing, it is flexible and long enough to be threaded joint-free where the use of fittings might be difficult.

1-7 Two types of Genova fixture supply valves, angled (left, No. 530651) and straight (right, No. 530301), let fixtures be served from the wall or floor. Both hand-tighten onto any 1/2″ supply pipe and accept 3/8″ O.D. riser tubes directly. The riser tubes reach up to the fixture.

conventional shower head with a water-saving one can save the average family more than 20,000 gallons of water a year, plus the cost of heating it to a comfortable showering temperature.

Similarly water-saving washbasin and sink faucets reduce the flow rate to 2.75 gallons per minute at normal water pressures. Toilets that are designed to be water saving will flush completely on only 1.5 gallons of water. Plumbing codes often require the use of water-saving, energy-saving devices.

Noisy plumbing
All plumbing should be designed and installed to perform quietly. One cause of plumbing noises is excessive water pressure. This can be solved by adding a pressure regulator to the system. It should be set at 50 psi. The regulator goes near the house main shutoff valve.

Sometimes metal piping systems are sound insulated by using cushioned mountings. This is not necessary with vinyl piping, as the thick thermoplastic walls minimize any water sounds.

Water hammer. One annoying plumbing noise is called water hammer. Houses in which the plumbing pounds angrily whenever a faucet or valve is quickly closed have this potentially destructive problem. Water hammer can result in considerable damage to a home's hot and cold water supply piping. The noise is caused by the inertia of rapidly moving but incompressible water in a piping system being brought to a sudden stop.

To prevent water hammer, typical plumbing codes call for the use of water-cushioning devices at most fixtures and appliances. Tests

1-8 These Genova Water Hammer Mufflers stop annoying water hammer. This occurs when rapidly moving but incompressible water is brought to a sudden stop by closing a valve.

in the Genova R&D lab have shown that the usual capped-pipe air chambers soon become waterlogged. In fact, before the first 200 shutoff cycles, their trapped air has largely dissipated, being replaced by water. No air, no cushion. Water hammer returns to the system.

This explains why so many homes suffer from water hammer, even though air chambers have been used.

What does work is Genova's Water Hammer Muffler (Genova Part No. 530901). Its air tight CPVC bulb surrounds an elastomeric bladder. The thick, pleated, balloonlike bladder separates air and water, thus preventing waterlogging. In addition, the space between bulb and bladder is factory pressurized. Like increasing the air pressure in your car's tires for hauling heavy loads, this makes the muffler more effective against overpressures.

When a faucet or valve closes quickly, instead of crashing to a hard stop, the flowing water diverts gently into the bladder. The bladder

balloons against internal pressurization, dissipating the flowing water's inertia.

Without an effective cushion, fast faucet shutoff is like crashing a car into a concrete wall. With one, it's like being stopped gently by a wall of marshmallows. Water hammer is gone. In fact, the Genova lab's tests showed that pipe-breaking overpressures of more than 600 psi are cushioned to slightly more than 100 psi by the Water Hammer Muffler. My own tests show a super soft shutoff, no matter how quickly the valve closes. Truly amazing!

A push on, hand-tighten adapter at the muffler's open end lets it be easily installed to any cold or hot water supply system.

Drain-Waste-Vent system

The Drain-Waste-Vent system (DWV) begins at the fixtures and appliances where the water supply system ends. Since drained water is not under pressure, the DWV piping must slope in the direction of desired flow. Water then flows by gravity from each fixture and appliance to the city sewer or to a private sewage disposal system.

Drainage fittings. Fittings used to carry away waste water are specifically designed ones called drainage fittings. They're made with

smooth, gentle bends on the inner surfaces (Fig. 1-9). This is to reduce flow resistance and also helps solids pass through. Drainage fittings are designed to have an included angle of slightly more than 90° to help the plumber maintain proper pipe slope.

Waste pipes. The DWV pipes leading away from all fixtures except toilets are called waste pipes. Toilets have outflow pipes, too, but they need such large ones that these are called **soil pipes**, not waste pipes.

Stacks. Usually fixture waste pipes and toilet soil pipes eventually empty into a vertical pipe that's called a stack. At its lower end, a stack leads into another run of pipes called the building drain. At its upper end, a vent stack is open to the air above the house roof. If it serves a toilet, a stack is called a main or soil stack. Every house has at least one main stack, measuring 3" or 4" in diameter. If a stack does not serve a toilet, it's called a secondary or waste stack. Then it may be only 1½" or 2" in diameter.

Building drain. The building drain is a horizontal pipe which is 3" or 4" in diameter and collects waste water from all the fixtures, including toilets, and leads it out of the house below ground. Once outside, 5 feet away from the foundation, a building drain becomes the house sewer.

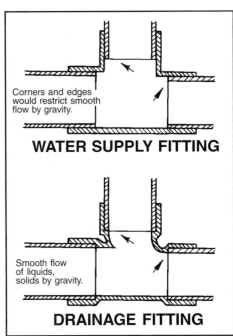

WATER SUPPLY FITTING

DRAINAGE FITTING

1-9

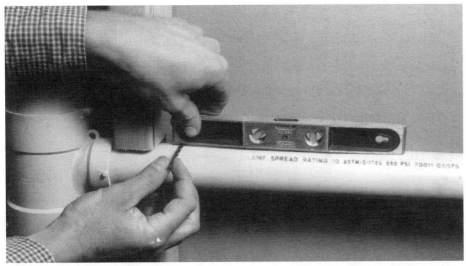

1-10

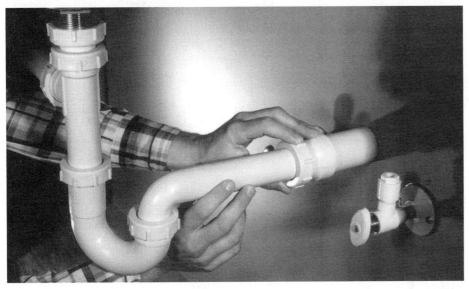

1-11

1-10 Drainage pipes should slope slightly in the direction of desired flow. Here a 3/16″ drill bit placed under the low end of a short 9″ level centers the bubble, indicating an ideal ¼″ per foot slope.

1-11 Every fixture needs a trap to prevent DWV and sewer gases from backing up into the house, yet let wastes flow out. Genova polypropylene trap is a lasting replacement for an old, corroded trap.

The house sewer pipe, usually 4″ in diameter, slopes beneath the ground toward a city sewer or private septic system.

Traps. Every fixture needs a trap. A trap is a U-shaped water filled pipe that allows water and wastes to pass through on their way out of the house, but prevents gases and vermin inside the DWV system from entering the house (see Fig. 1-11). Toilets contain built-in traps; other fixtures use separate ones.

Access must be provided for cleaning traps, should they become clogged. Toilet and sink/washbasin traps can be cleaned by working through the fixture drain. Some bathtub and shower traps are cleaned through the drain, as well. Those that cannot be reached from the drain must have access for cleaning through a hole in the floor or working from a basement or attic

crawlspace. Some traps have cleaning/draining openings of their own.

Venting. The need for traps brings on still another necessity—venting. As you may know, water rushing down a pipe by gravity creates a vacuum or suction at the high end of the pipe. This is called **siphon action.** Siphon action is powerful enough to let atmospheric pressure force all the water out of a trap, leaving it nearly dry. Venting to outside air above the roof equalizes the pressure on both sides of the trap and prevents trap siphoning (Fig. 1-14). Inside-the-house venting works, too, if sewer gases are kept from escaping. A NovaVent™ Automatic Anti-siphon Valve (Genova Part No. 740151 and 740401—see Ch.14) does this, letting air enter the DWV system under a vacuum to break up siphon action, but not letting gases out under

pressure. The larger NovaVent™ Relief Valve (Part No. 740401) can be used to vent an entire bathroom, including the toilet, as an alternative to running a vent pipe up through the roof.

Atmospheric venting also prevents pressurization in the DWV system or sewer system from reaching the point where it could force past a trap's water seal.

Every trap in a plumbing system, including the one built into the toilet bowl, must be vented.

Branch venting. If the waste pipe from a fixture is short enough or large enough, a fixture's trap can be both drained and vented through the same pipe. The length of pipe from the trap to its vent is called the **trap arm.** Trap arm length is limited by code and by what will work without trap siphoning. All fixtures are both drained and vented for a short distance through their trap arms, but if a vent stack isn't close behind the fixture, its trap arm would be too long. Then alternate venting must be provided. This may be in the form of

How a trap works

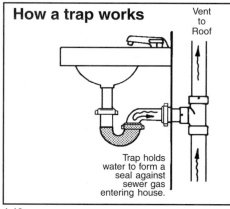

Vent to Roof

Trap holds water to form a seal against sewer gas entering house.

1-12

How a toilet works

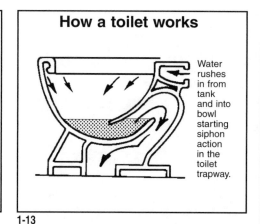

Water rushes in from tank and into bowl starting siphon action in the toilet trapway.

1-13

How a vent works

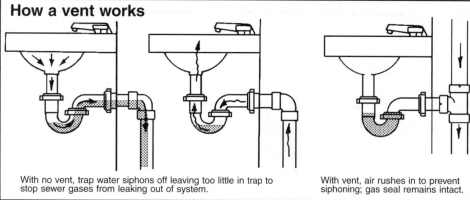
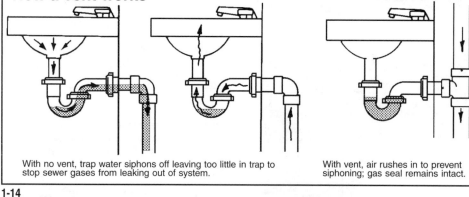

With no vent, trap water siphons off leaving too little in trap to stop sewer gases from leaking out of system.

With vent, air rushes in to prevent siphoning; gas seal remains intact.

1-14

Trap-Arm Venting

Stack

Lavatory

Wet Vents

Toilet

Bathtub

Branch Vents

Stack

Branch Vents Added

Toilet

Bathtub

Lavatory

1-15

1-16

1-16 By devoting a few weekends, you can completely remodel a tired, old bathroom into exactly what you want. The tearing-out portion of the job is quickly completed, leaving the pleasant plumbing and finishing to look forward to.

another vent stack closer to the fixture or a branch vent which is sometimes called a "revent." This can also be done with a NovaVent™.

Stacks, vents, and NovaVents™ are usually made from the same size pipe that is used for draining the fixture. Sometimes, vent pipe that is a size or two smaller, is permitted by code. In any case, no drain-waste-vent pipe is smaller than 1¼". In modern plumbing, 1½" is the smallest size pipe ordinarily used.

Cleanouts. Another DWV system necessity is for cleanout openings, which permit the removal of inside-the-system stoppages, should any occur. A horizontal drain run must have access for cleaning it. This also

includes sewer pipes. Cleanouts are often made by locating a cleanout opening at the high end of the run and plugging it. Removal of this plug allows access for drain cleaning. Cleanout openings must be accessible and should also enter in the direction of flow (when you look into them, wastes flow away from you).

While cleanouts are sometimes made with special cleanout fittings, Genova's Twist-Lok™ plug works with any same-sized plastic fitting. Almost every Genova DWV fitting will accept the soft gasketed Twist-Lok™ plugs. The on-off connection is made easily by hand and there are no threads to corrode and make

cleanout plug removal difficult.

One more thing—as has been mentioned, because a Genova plumbing system is nonmetallic, it cannot conduct wayward electricity, such as lightning, to your plumbing fixtures. And it does not need to be bonded into the house electrical system. Do not depend on your Genova plumbing system to provide electrical pathways as it does not carry current.

Those are the basics of the home plumbing system. The following chapters tell how you can do your own plumbing the easy way using Genova pipes and fittings.

Chapter 2 Plumbing Codes and Inspections

2-1

2-2

2-3

2-1 All piping systems, including sewers, will likely have to be tested for leaks. And the inspector will want to witness part of the test. Here a water-soluble dye is poured into a plugged sewer's cleanout opening while it is filled with water. Any leaks will show up downstream as colored water.

2-2 To fill-test the DWV system for code approval, all openings need to be sealed off. This is taken care of by Genova's handy "Pop-Top" toilet flange. It contains a vinyl knockout shield that closes off the toilet's soil pipe. The shield also helps to keep debris out of the DWV system. Just before the toilet is installed, the shield is tapped out with a hammer (while wearing eye protection).

2-3 The lower end of the building drain is capped off for a water-fill test. Later, the DWV system can be drained without removing the cap by removing a sheet metal screw threaded into a drilled hole. This way, the DWV system may be checked periodically as it goes up. The final test is a fill to overflowing at the roof vents.

When you read about doing plumbing projects, you often see the words, "Check your local code." This is because the local code is a set of rules you are obliged to follow when installing plumbing. It's likely that the code is based on one of the model plumbing codes in the United States. None of these prevents you from doing your own plumbing, though.

Genova products are accepted by most codes. Most Genova Products also meet Federal Housing Administration (FHA) and National Sanitation Foundation (NSF) requirements.

A word about codes. Plumbing codes are designed to protect the public by setting minimum requirements for safe installation. All codes are hard-put to keep up with rapid changes in technology. For example, new materials are constantly being developed that didn't exist when the codes were written.

Plumbing inspector. If you do code-covered plumbing on your house, it will most likely need to be inspected. The person who does this is the local plumbing inspector. The usually helpful inspector will also want to witness tests of various parts of your system (see Figs. 2-1 to 2-7). These tests are described in later chapters.

Plumbing inspections and tests usually come at three stages of new

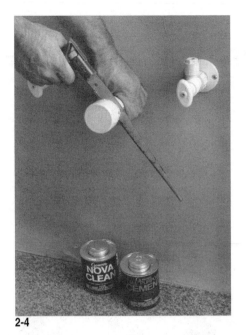

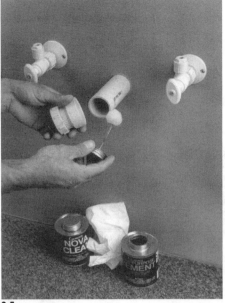

2-4

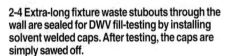

2-5

2-6

2-4 Extra-long fixture waste stubouts through the wall are sealed for DWV fill-testing by installing solvent welded caps. After testing, the caps are simply sawed off.

2-5 Then the pipe end and fitting socket are prepared for solvent welding.

2-6 Finally, a Genova Part No. 72211 trap adapter is solvent welded onto the stubout. During finish plumbing the adapter will accept a tubular fixture trap.

2-7 Once your piping has passed the no-leaks test and your plumbing inspector has okayed its compliance with the code, you can confidently close in the walls. The rough plumbing has been completed.

2-7

construction: (1) rough-in, (2) sewer, and (3) final. On visits, the inspector will check to see that your piping materials and workmanship conform to local requirements.

It's the plumber's—YOUR—duty to let the inspector know when the project is ready for an inspection. The inspector will describe the exact condition wanted.

The plumbing inspector is often the bright side of the picture. At the least, he or she is a good person to have on your side. The inspector is authorized to interpret the local plumbing code and, if the inspector wishes, some of the rigidity can be removed from it to make the code better fit you as an owner-plumber. What's more, nearly every inspector will go out of the way to explain things to a do-it-yourself plumber, provided the knowledge is earnestly sought.

Your goal in all dealings with the inspector should be to make the official want to help you. A warm greeting followed by good treatment is the order of the day. Seek advice. Plumbing inspectors come in all shapes and sizes, but there is no substitute for one who wants to help you. One builder I know advises handling an inspector like cooking a fish—lightly and gently.

Common code concerns

Plumbing codes come out strongly in a number of areas. These are the ones that will draw the most attention from your plumbing inspector.

The number one code concern is any cross connection between the water supply and the drainage system. These are not permitted by ANY code.

Whether for DWV or pressurized water supply piping, proper pipe sizing is important, too. Codes contain pipe sizing tables for both

water supply and drainage pipes.

Because it depends on gravity flow, drainage piping gets lots of code attention. All drain pipes must be properly sloped. Also, codes don't like abrupt 90° changes in flow direction of wastes, as these tend to cause blockages. Although some 90° turns are permitted, the use of long-turn 90° bends and pairs of 45° turns to accomplish a 90° turn are preferred for drain runs.

There are limits on the lengths of trap arms (Figs. 2-8, 2-9). Double-trapping of fixtures is not allowed. Moreover, traps may not be larger than their drains.

The way a vent stack ends above the roof is a common code call. To prevent ice closure in very cold climates, some vent terminations must be larger than the vent pipes themselves, via a vent-increaser. This is made with pipe reducers accommodating larger pipe sizes. Proper

Trap Arm (shaded portion A&B)

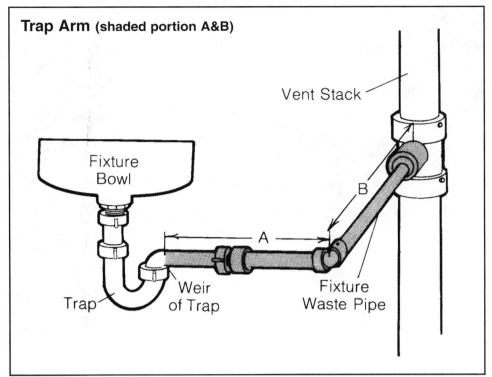

Vent Stack

Fixture
Bowl

B

A

Trap

Weir
of Trap

Fixture
Waste Pipe

2-8

2-8 Trap Arm (Shaded Portion—A +B)

2-9 The length of trap arms is limited by plumbing codes. Often the limitation is much shorter than actual tests indicate will work. While this 1½" washbasin waste pipe could reach up to 8' and still function well, codes limit its length to as little as 3½'.

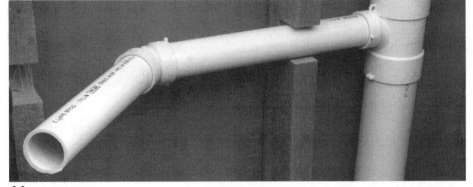

2-9

pipe support is called for by the code, too. And don't be confused by codes that refer to toilets as **water closets.**

Working under codes

Codes, while they are not law, carry the force of law. This means that, should you exercise the right to do your own plumbing free from regulation, you face the police power of local government. It can be awesome. The good news is that Genova's plumbing materials are widely approved by up-to-date plumbing codes around the world and where applicable they conform to standards of the renowned American Society for Testing and Materials.

If it's typical, the code will let you choose among safe, proven piping materials, to be installed in a safe-for-the-community manner widely recognized by plumbing authorities. However, a few localities still have not gotten around to okaying the use

of the most up-to-date piping materials. If you live in one of these areas, here's how to work under such a code.

First of all, a good deal of home plumbing can be done without invoking the code. What usually triggers code inspections and an official's resultant insistence on code compliance is a building permit.

Often, the need for a building permit depends simply on what you call your project.

New construction, for example, just about always requires a building permit.

Maintenance and repair—replacing old pipes with new ones, for instance—almost never requires a permit.

Remodeling may or may not need a permit, depending on what's to be done. Simply modernizing your plumbing fixtures without changing their locations should be permit free.

On the other hand, relocating of plumbing fixtures or converting a large closet into an add-on bathroom probably would need a permit.

To find out, check with your local building department. See whether a permit is necessary for what you wish to do. Then perhaps you can avoid getting one.

While you're at it, get a copy of the local plumbing code. Read it and interpret the code's provisions as they apply to your project. Then proceed in the manner you judge best.

2-10 Don't cover up your piping below ground or behind walls until it has been checked and approved by the local plumbing inspector. This is because most local codes require inspections, and the plumbing inspector will want to look at everything.

Getting a Variance

If you have trouble with an overly restrictive code, stand up for your rights. Ask the chief plumbing inspector for what's called a **variance.** It's simply a permitted deviation from the code. A variance would never be issued to allow unsafe conditions, however, one might let you use a piping material that is widely okayed by other codes yet not currently accepted locally. Tell the official that you'll be happy to make your project a local trial of what's done elsewhere. Any plumbing official who refuses even to learn more about new, accepted plumbing materials is skating on very thin ice.

So whenever you read, "Check your local code," don't panic. A code is just a book of rules with a friendly inspector who comes along to help explain them. If you'll treat it that way, you'll do fine at home plumbing under a local code.

Section 2 Doing Rough Plumbing
Chapter 3 Solvent Welding

3-1

3-2

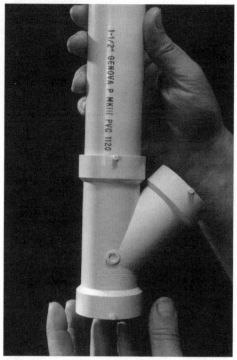

3-3

Of the various methods for joining pipe, tubes, and their fittings, it's tough to beat solvent welding for low cost and ease. Solvent welding is just about foolproof. Safe, too, since there's no flame to start a fire in your house while the water is turned off. A fine-tooth saw, a knife, and sometimes a brush are the only tools needed.

The solvent welding plastic pipes are comprised of PVC, CPVC, ABS, and styrene. PE, PEX and PP pipes cannot be solvent welded, as they are resistant to solvents. Solvent cement is comprised of a plastic filler dissolved in a mixture of active solvents. Sometimes the mixture is clear, sometimes pigmented. Clear or pigmented, the action is identical. The solvent welded pipe and fitting are permanently joined into a single unit.

Fit of Parts

Currently manufactured pipes and fittings for solvent welding are designed to have what's called an **interference** fit. This makes space for the solvent cement, yet leaves no voids after the solvents have evaporated. Sometimes, however, a poor fit can be more common than a good fit. When a pipe by one manufacturer and a fitting by another

are used together, it's easy for a joint to be far too loose for successful solvent welding. The happy side is that you can quickly check for proper fit before making up a solvent welded joint. Do it before applying any solvent cement.

First check to see that the pipe will enter the fitting (Fig. 3-2). And you should be able to feel moderate resistance for one-half to two-thirds of the socket depth. This resistance will keep the fitting from falling off when turned upside down (Fig. 3-3).

Even though the fit is extra tight, if the pipe can be started into the fitting, you can still solvent weld it successfully. But, in that case, be sure that the pipe end gets a second coat of solvent cement before insertion (see Fig. 3-10).

Poor fit. If the fit is much too loose, don't proceed with solvent welding. The best solution is to insist on getting pipes and fittings of the same brand. Even so, it's still good solvent welding practice to test-fit each joint. Genova is the only brand where you can safely skip test-fitting.

The reason for all the care is that when you close your plumbing in behind the wall, you don't want to risk any leaks.

3-1 Solvent welding goes so easily that it's tempting to skip a step. Don't. If you do it right, it's almost impossible to get a leak.

3-2 Pipes and fittings for successful solvent welding need an interference fit. If you stick with Genova pipes and fittings, you can be sure the parts fit together properly. When using other brands, be sure to check for proper fit before solvent welding them. The pipe should enter the fitting, yet there should be moderate resistance for one-half to two-thirds of socket depth.

3-3 The fit should be tight enough that the fitting will not fall off when the pipe is inverted. If the fit is too loose, the joint cannot be solvent welded successfully.

Making the weld

To get a joint ready for solvent welding, cut the pipe off squarely. Use a plastic tubing cutter (Genova Part No. 534991) on tubes up to 3/4". On larger tubes and pipes, use a fine-tooth saw or large-wheeled pipe-cutter (see Figs. 3-4, 3-5). To get a square end when sawing pipes, do it in a miter box, if you have one. Remove any burrs as illustrated in Fig. 3-6, and it's best to chamfer the ends (Fig. 3-7). Inspect pipe ends,

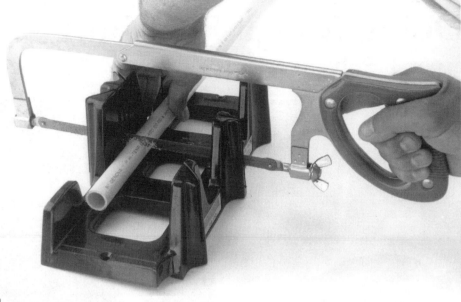

3-4

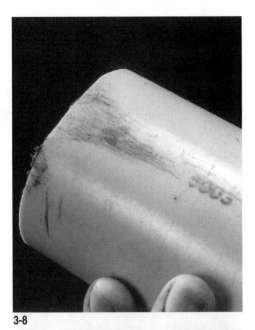

3-8

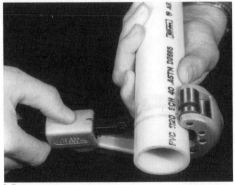

3-5

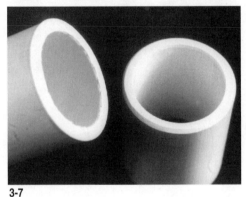

3-6

3-7

3-4 The first step in successful solvent welding is to cut off the pipe end squarely. Any fine-toothed saw—a hacksaw—works fine.

3-5 A pipe cutter with larger plastic-cutting wheel ensures square cuts. An off-square cut minimizes the depth of the solvent welded joint.

3-6 Remove all burrs on the pipe end using sandpaper or a knife. At the same time, check the pipe-fitting mating parts for gouges, deep abrasions, and cracks.

3-7 The pipe at the left was cut by a pipe-cutter. At right, the inner and outer edges have been beveled, a nice touch that's done before solvent welding.

3-8 When a pipe has been dragged on its end, flat wear spots appear. These should be cut off. It's better to throw away an inch of pipe than to risk a leak.

cutting off bad ones (see Fig. 3-8).

If the pipe ends are oily, dirty, or greasy, it is best to use the two-step solvent welding process. Some codes require it. In two-step solvent welding, the pipe and fitting surfaces to be joined are first treated with Novaclean®, a high-quality all-purpose cleaner/primer. Novaclean® not only removes joint-spoiling dirt, oil, and grease, it etches the mating surfaces to open them to solvent attack (Fig. 3-9). Apply Novaclean® with a clean, lint-free cloth. Follow by applying the solvent cement onto the cleaned and dry surfaces.

The two-step cleaner/primer procedure, while not mandatory, is sometimes inexpensive insurance for leak-proof joints. It's like omitting the use of flux when sweat soldering copper tubing. While you might be able to get away with it sometimes, the few leaks you'd get would take longer to fix than to have used the flux (or Novaclean®) in the first place. That's why Genova recommends two-step solvent welding.

How to solvent weld a pipe-fitting joint is shown in Figs. 3-9 through

3-12. Be sure to follow any specific instructions on the can of solvent cement, because once assembled, the joint is permanent. There's no way to back up and perform a missing step.

Cautions. The biggest single cause of joint failure is inadequate coating of the joining surfaces with cement. The sequence is important, too. You want any excess cement to bleed to the outside of the joint when the pipe is pushed into the fitting. By applying only one coat to the fitting socket, there is less chance of having too much cement forced into the bore of the pipe or fitting to cut down on water flow.

Make sure that all solvent weld joints are dry, and keep moisture away from your solvent cement as well. Water or dirt that gets on a joint after assembly won't hurt.

The resulting joint can be handled gently within a minute, take water flow within half an hour, and handle full water pressure in two hours (depending on temperature, humidity, and the solvent cement being used).

3-9

3-10

3-11

3-12

3-13

A full cure takes about 24 hours. During that time the solvents evaporate, leaving only pure plastic resin in the joint. It becomes one with the pipe and fitting, thus the term **solvent welding** (Fig. 3-16).

Errors. If you make an error in solvent welding, cut out the wrong fitting and install the correct one, using couplings and pipes to fit it in. (This is shown in Fig. 7-27.)

About solvent cements

Genova solvent cements are the highest quality cements you can buy. Each contains lots of a costly chemical—tetrahydrofuran. This is what makes them work so well. Some manufacturers skimp on such high-cost ingredients.

Genova has a number of cements: Novaweld® C, Novaweld® P, Arcticweld® P, ABS, and All Purpose

Cement. They come in 1/4 pint, 1/2 pint, pint, and quart sizes. For how much cement you need to do the job, see Table 3-A. For the two step method, usually, the same amount of Novaclean® is required.

The solvent cement you use should always be matched to the type of plastic you're working with: PVC, CPVC, ABS, or styrene (Table 3-B). Check the label. Genova All Purpose Cement works well with all of these. ASTM specification F493 sets out the requirements and cementing procedures for solvent cements. Each Genova solvent cement meets these, plus the specific ASTM requirements for its own type of cement. For example, Novaweld® P meets ASTM D2564. Labeling of Genova cements conforms to Consumer Protection Agency requirements.

3-9 Begin the solvent welding process by checking the pipe end for damage, dirt, oil, and grease.

3-10 Then apply solvent cement (matched to type of plastic being welded) to the pipe end.

3-11 Apply cement to the fitting socket. If you wish, the pipe may be given a second application for insurance.

3-12 Without waiting, join the pipe and fitting with a slight twist, bringing the fitting into final alignment. It's a good idea to hold the parts together for a few seconds, letting the solvent cement set enough to hold them. Once the joint has set a little, alignment cannot be changed.

3-13 Smaller CPVC water-supply tubes and fittings are joined similarly by applying Genova's All Purpose Cement. This high-quality cement alone will handle whole-house plumbing.

Table 3-A
PVC Solvent Cement Requirements

Fitting and pipe size	Average number of joints solvent welded per can	
	1/4 pint	1 pint
1/2"	63	255
3/4"	42	170
1½"	17	68
2'	9	38
3'	7	30
4'	5	21

Table 3-B

Material	Novaclean® All-Purpose Cleaner	All-Purpose Cement	Novaweld® C	Novaweld® P	Arcticweld® P
ABS	R	R	R	R	R
CPVC	R	R	R	NR	NR
PVC	R	R	R	R	R
Styrene	R	R	R	R	R
PP and PE	NR	NR	NR	NR	NR

R=Recommended   NR= Not Recommended

3-14

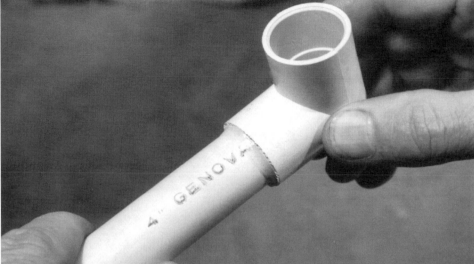

3-15

3-14 Cement applicator size should correspond with the size of piping being joined. Ideal is an applicator that's one-half of pipe diameter. If the applicator that comes with a cement is too small for the job you are doing, you can use a paint brush of the correct size.

3-15 If you can see a fillet of solvent all around the fitting, you know that enough solvent cement was used. In practice, it's hard to use too much cement.

3-16 A solvent welded plastic pipe and fitting, as this cross-section shows, are joined into a continuous unit. No weak spots are left for corrosive attack or for pipe-fitting separation.

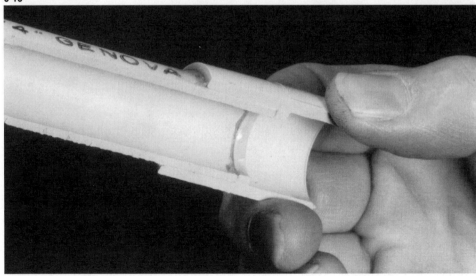

3-16

Do your solvent welding of larger DWV pipe sizes across a pair of 2" x 4"'s to keep the pipes and fittings aligned and off the floor. Take care not to spill solvent cements on a finished floor.

Temperature. Most solvent cements are formulated to work best at or near room temperature. If used where it's really cold, they tend to set too slowly. However, a normal cement may be used beyond its limits if you allow for the consequences. For example, in hot weather, you can work more quickly; in cold weather you can give more hold-together time and more time for curing. That way, one cement can handle all temperatures. Nevertheless, working with PVC at temperatures below 40°F, a low-temperature cement such as Genova's Arcticweld® P may prove helpful.

Precautions. When handling solvent cement, it's best to work in a well ventilated space: leave a door open, a window, too, if possible. Cap the cans between uses and keep cements and cleaners away from an open flame. Don't smoke. The precautions are spelled out on the can labels. Solvent cement that has dried on your hands can be removed with a good mechanic's hand-cleaner.

Aging of cement. Evaporation makes a solvent cement thicken in the can. Even though thickened cement may still be able to be used, it may have lost some of its effectiveness. In any case, don't try to thin a solvent cement. Moreover, never mix old, thickened cement with fresh.

3-17

3-18

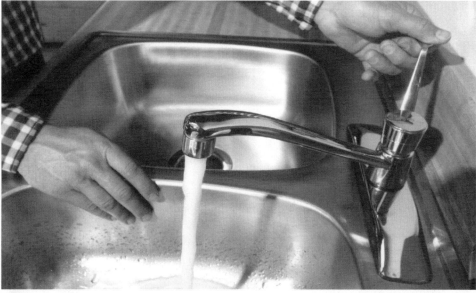

3-19

3-17 The solvent cement should always be matched to the type of plastic being solvent welded. Left to right, Genova Novaclean®, if you wish may be used on all materials as an all-purpose cleaner/primer; Genova All Purpose Cement works on all four solvent welding plastic piping materials; Novaweld® C is intended specifically for use with CPVC, but also may be used on all solvent weldable plastics; Novaweld® P is for use with PVC, ABS, and styrene. Arcticweld® P is for use in cold temperatures.

3-18 Genova markets a super-handy solvent welding kit that's ideal for small jobs. The kit contains a 1 oz. bottle of Novaclean® and a same-sized bottle of Genova All Purpose Cement with self-contained dauber. Instructions for use are printed on the package. It's Genova Part No. 141001.

3-19 Flushing out the CPVC water supply system gets rid of all traces of solvent cement and any debris in the system. Thus the water you draw will be pure and potable from the outset.

Cements that turn yellow are getting old. Yellowing isn't harmful as long as the cement will pour. If it turns rusty or dark brown, it means that it is too old for use and it should be thrown away.

Flushing the Water Supply System
In a solvent welded CPVC water supply system, it's a good idea to flush out the piping before pressurizing it. Flushing also expedites the curing of the solvent cement and helps to purge the system of tastes and odors.

To do this, first close the house main shutoff valve. Then fully open all outlets, such as faucets. Go back and open the house main valve just enough to let water flow slowly into the new system. When water reaches the lowest opening, it will begin flowing out there. Close this valve off to a steady trickle. Move up through the house, closing down each flowing outlet as you come to it. Let the water trickle from each for 10 minutes. This will carry away any solvent vapors left inside the system-- the vapors actually dissolve in water and flow away.

Then you can close off the main valve completely for half an hour, letting water stand in the system. After that, reopen the main shutoff for another 10 minutes of trickling. Repeat the cycle one more time.

Finally, pressurize the system by turning the main valve on full and flush out any debris from each faucet by opening it fully for a few seconds (Fig. 3-19). Then close it. Don't forget to flush each toilet, too.

"Vinyl plumbing pipe is so chemically

resistant it's used in chemical plants to convey concentrated acids and strong alkaline solutions."

Chapter 4 Vinyl DWV Pipes and Fittings

4-1

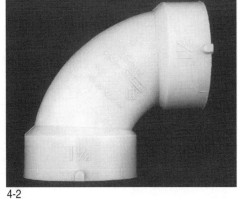

4-2

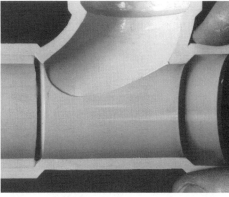

4-3

4-4

4-5

There's no need for you to avoid doing Drain-Waste-Vent plumbing simply because the pipes are large. Using the Genova PVC-DWV system, they go together almost as easily as smaller water supply pipe and fittings. Vinyl, which is short for polyvinyl chloride (or chlorinated polyvinyl chloride), excels at DWV service. Vinyl piping materials have high chemical and corrosion resistance. In fact vinyl pipe is so chemically resistant it's used in chemical plants to convey concentrated acids and strong alkaline solutions. These would quickly destroy metal piping. Genova vinyl pipes are unaffected by salt solutions, alcohols, and many other chemicals. You can pour drain-cleaning chemicals down your PVC drains with full assurance that the pipes are completely impervious to them.

A PVC system will also take all the hot drain water you could want to put down it.

The makers of ordinary "plastic" plumbing products cannot match this. With Genova PVC-DWV, you get chemical plant quality plumbing for your home.

Unlike black ABS piping, PVC does not suffer stress cracking in the presence of fats, greases, and acids. Furthermore, Genova vinyl pipe is light, a 10-foot length of 2″ Genova PVC-DWV pipe weighs only six pounds.

It's highly impact resistant, too. Hammer blows that would hopelessly dent copper pipe merely glance harmlessly off Genova rigid vinyl pipes. Rough handling, back-filling, traffic loads, temperature changes, vibration—all hazards that make ordinary piping systems unusable—are no problem with Genova's superior vinyl pipe. What's more, with Genova PVC there's no possibility of electrolysis eating away at your DWV system. And because of the high quality resins used, Genova rigid vinyl possesses outstanding weathering characteristics. Moreover, Genova vinyl DWV products are fire-resistant. In fact, they have been awarded a flame-spread rating of 10, in accordance with ASTM procedure E-84. That's good!

The interior passages of Genova PVC-DWV pipe offer less

4-1 Avoid this! When you buy drain-waste-vent pipes and fittings, be sure to "squeeze the bananas"; that is, examine each for quality. For example, don't choose a deformed DWV pipe when you can just as easily buy a straight one that will ensure good gravity flow. Buying Genova DWV brand identified products is one assurance of quality.

4-2 Drainage fittings are designed to make gentle bends, allowing wastes to flow by gravity without clogging. 90° turns should be avoided, unless long-sweep elbows are used. Otherwise, use two 45° elbows to accomplish a 90° turn.

4-3 The insides of a drainage tee (shown here in cross-section) feature smooth, catch-free passages for waste flow by gravity.

4-4 While you're shopping for quality plastic DWV products, shun those with excessive "flash" or fins that stick out. While this exterior flash would not harm fitting flow characteristics, it indicates a sad lack of quality control in manufacture. Look for another brand.

4-5 The inside of this drainage fitting presents a house of horrors passage for flowing wastes. The ridges and grooves left by a worn-out injection-mold core create snags that would encourage clogging. Fittings like this should have been rejected before they reached your store's shelves. Avoid buying that brand.

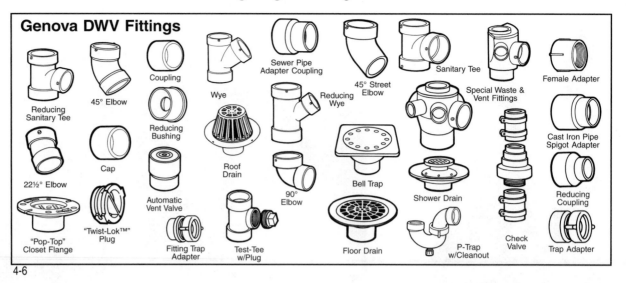

Genova DWV Fittings

Reducing Sanitary Tee · 45° Elbow · Coupling · Wye · Sewer Pipe Adapter Coupling · 45° Street Elbow · Sanitary Tee · Female Adapter · 22½° Elbow · Cap · Reducing Bushing · Reducing Wye · Special Waste & Vent Fittings · Cast Iron Pipe Spigot Adapter · "Pop-Top" Closet Flange · "Twist-Lok™" Plug · Automatic Vent Valve · Roof Drain · Bell Trap · Shower Drain · Reducing Coupling · Fitting Trap Adapter · Test-Tee w/Plug · 90° Elbow · Floor Drain · P-Trap w/Cleanout · Check Valve · Trap Adapter

4-6

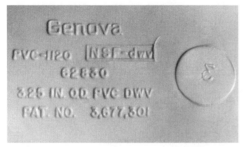

4-7

4-7 This closeup shows the markings identifying a Genova Schedule 30 In-Wall vinyl fitting for DWV use and indicating its acceptance by the National Sanitation Foundation for DWV use. This is Part No. 62830, a 3" Schedule 30 In-Wall 90° elbow.

4-8 The chemical resistance of Genova PVC pipe (top) is much greater than that of ordinary ABS plastic pipe (bottom). Both pipe ends were doped with PVC solvent cement, then scratched. The excessively softened ABS scratched much more deeply. As you can see, Genova PVC is the superior piping material.

4-9 A 3" Schedule 40 vent stack running up the wall behind a toilet calls for the 2" x 4" stud wall to be furred out almost an inch with wood strips nailed to the studs. The use of Schedule 30 In-Wall PVC-DWV pipes and fittings makes this step unnecessary.

4-8

4-9

resistance to flow, resulting in higher through-put, both initially and after years of service. Surfaces stay smooth, DWV lines stay open.

But the best part is that Genova PVC installs simply using solvent welded joints. So you can see that Genova's PVC is clearly the material of choice for making a DWV system.

Genova DWV pipes and fittings (Fig. 4-6) come in 10-foot lengths and five diameters: 6", 4", 3", 2" and 1½".

Schedule 30 In-Wall

Genova's DWV products are available in 700 Series Schedule 40 PVC-DWV and in 600 Series Schedule 30 In-Wall PVC-DWV. 700 Series Schedule 40 Genova pipes and fittings meet or exceed the requirements of ASTM D2665. Genova 600 Series Schedule 30 In-Wall pipes and fittings conform to ASTM D2949. The **In-Wall** designation

by Genova comes from the fact that everything—fittings as well as pipe—fits within a standard 2" x 4" stud wall. Schedule 30 In-Wall was developed specifically by Genova for do-it-yourself home plumbing.

The whole idea of Schedule 30 is to get around having to fur out 2" x 4" stud walls to accommodate 3" pipes and fittings (Fig. 4-9). In-Wall pipes are the same size inside as 3" Schedule 40 pipes. But their somewhat thinner walls make them smaller on the outside. While a 3" Schedule 40 DWV pipe will just fit inside a 2" x 4" stud wall, its fitting hubs will not. But with Schedule 30 In-Wall, the hubs will fit also (Fig. 4-10). Thus, with Schedule 30 In-Wall, no furring-out is necessary. Like Schedule 40, Schedule 30 In-Wall may be buried in the ground (check local code).

Only Genova offers a complete Schedule 30 product line. Standard

4-10

4-11

4-10 Genova Schedule 30 In-Wall pipe and fitting (left) will just fit inside a 2" x 4" stud wall, while the Schedule 40 fitting at the right will not.

4-11 The difference between Schedule 40 (right) and Schedule 30 In-Wall is the thickness of the pipe and fitting walls. In-Wall, designed for the do-it-yourself home plumber, is accepted by almost all codes.

3" Schedule 40 pipe will not fit 3" Schedule 30 In-Wall fittings, so don't try to mix the two. However, adapters will let the two pipes be joined. My advice: avoid Schedule 40 in the 3" size. Use Schedule 30 In-Wall, instead.

Here's a handy recap for your Genova DWV ordering:

All 3": order Genova 600 Series Schedule 30 In-Wall pipe and fittings. Any other size-- 1½", 2", and 4" pipe --order Genova 700 Series Schedule 40. These will fit Schedule 30 In-Wall reducing fittings directly without adapters. For example, a 3" x 3" x 1½"

Schedule 30 In-Wall tee fits Schedule 30 In-Wall pipe at its larger 3" sockets and a 1½" Schedule 40 pipe at its smaller branch socket.

One more thing: Genova's Schedule 40 PVC pipe is dual rated for PVC-DWV systems and PVC Pressure systems. This means the pipe meets ASTM D2665 for DWV systems as well as ASTM D1785 for Pressure systems. Only the pipe is dual rated-- DWV and Pressure each have their own fittings.

4-12 Schedule 30 In-Wall pipe measures 3″ on the inside, about the same as a Schedule 40 pipe.

4-13 But because of its thinner walls, Schedule 30 In-Wall pipe measures only 3¼″ in outside diameter.

4-14 And a Schedule 30 In-Wall fitting's hub measures just 3⅝″ across-- slim enough for the fitting to go inside a 2″ x 4″ stud wall.

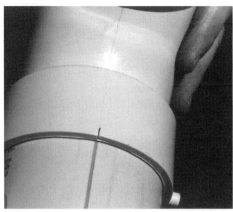

4-15

4-16

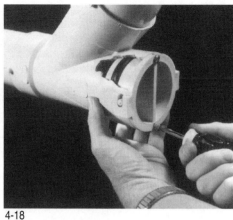

4-18

How the Genova plumb line/degree mark system works

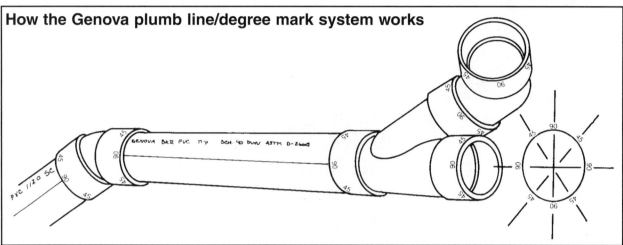

4-17

Unique features

Along with the quality of Genova pipe and fittings come some downright useful features to help you plumb.

Plumb lines. One such feature is Genova's unique Plumb Line/Degree Mark system (Fig. 4-15). Every Genova PVC-DWV pipe features a colored line running its full length. The line is a reference to help you keep pipe and fittings aligned as you assemble them. To eliminate guesswork with the plumb line, all angle-dependent Genova PVC-DWV fittings are degree marked around their edges. Arrows indicate 45° alignment; the projecting pins and mold marks indicate straight-on alignment. You only get it with Genova.

"Pop-Top." Genova "Pop-Top" toilet flanges are used to join toilets to a DWV system.

Twist-Lok™. Every horizontal run of drainage pipe should have a cleanout opening at the high end. There is no easier way to make this than with Genova's Twist-Lok™ plug. The Twist-Lok™ plug slips into a tee, wye, or other fitting socket, locking with a twist. Its slotted ears hook over the two pins found on some Genova

fitting hubs. Without tools, it allows almost any Genova DWV fitting to be used as a cleanout. A soft sponge rubber O-ring gasket seals tightly inside any same sized fitting hub, whether Genova or not (Fig. 4-18). Twist-Lok™ plugs also have special screw slots in their rims enabling them to be used on plastic DWV fittings which do not have pins. Appropriate sheet metal screws easily self tap between the Twist-Lok™ and the fitting hub assuring a secure fit.

Street/Socket Design. Genova reducing tappings in 3" and 4" fittings come sized for either 2" or 1½" pipes. This is Genova's unique Street/Socket Design (SSD) applied to Drain-Waste-Vent. The inlet accepts a 2" fitting over its outside diameter (Fig. 4-19) or a 1½" pipe into its inside diameter (Fig. 4-20). Street/Socket Design in DWV saves you having to order separate 1½" and 2" reducing fittings, or having to specify what reduced size you need, since Street/Socket Design works on both sizes.

A handy leaflet on PVC-DWV is available free from Genova as Part No. 252176.

4-15 Degree marks around the outsides of all Genova PVC-DWV directional fittings, together with a plumb line on every pipe, let you accurately align fittings as you solvent weld them together. Marks and projecting pins show the 45° and 90° positions.

4-16 Genova "Pop-Top" closet flange for toilet installation contains a flashed-over opening to keep the insides of the DWV system clean and to seal off the opening for water testing. When you're ready to install the toilet, simply tap the shield out with a hammer. Be sure to wear safety glasses.

4-17 How the Genova plumb line/degree mark system works.

4-18 This cross-sectioned photo shows a Genova Twist-Lok™ cleanout plug getting sheet metal screws between the plug and fitting hub before water-testing a multistory DWV system with high head pressure. The cleanout fits all standard plastic DWV hubs without the need for cleanout adapters. When used in non-Genova fitting hubs without pins, the fitting needs screws to hold it in place no matter how low the head pressures are during water-testing.

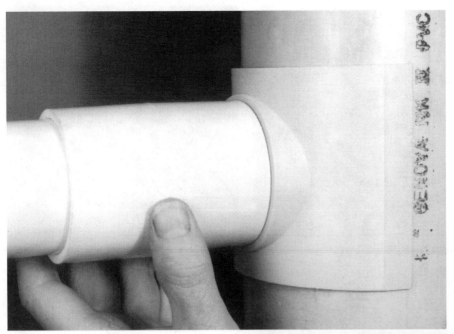

4-19

4-19 A handy Genova Street/Socket Design DWV fitting accepts a 2" DWV fitting over its outside diameter. The 2" coupling then accepts a 2" pipe to continue the run.

4-20 Or it accepts a 1½" DWV pipe on its inside diameter.

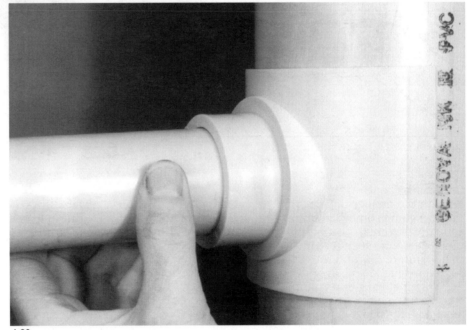

4-20

Table 4-A
ASTM DWV Standards
D2661 (ABS)
D2665 (PVC)

Pipe Dimensions and Tolerances

Nominal Size	Outside Diameter Average	Tolerance	Out-of Round	Wall Thickness Minimum	Tolerance
3"	3.500"	±0.008"	±0.015"	0.216"	+0.026" −0.000"

Fitting Dimensions and Tolerances

Nominal Size	Socket Entrance Size Average	Tolerance	Socket Bottom Diameter Average	Tolerance	Socket Depth	Min. Wall Thickness
3"	3.520"	±0.015"	3.495"	±0.015"	1½"	0.219"

Facts on fittings

There's a whole story to be told on buying DWV pipes and fittings. It's summed up: get everything from one manufacturer...and that manufacturer should be Genova. Here's why:

ASTM standards for DWV pipe and fittings represent the consensus of pipe and fittings manufacturers' thoughts on the dimensions they can economically hold to on a mass production basis. Nothing wrong with that. However, those dimensions and the tolerances allowed in them were based on the fact that most manufacturers made pipes only, or fittings only —not both, as Genova does.

Take an example, a 3" size, say, and see how the standards actually work out in the real world. From Table 4-A showing ASTM pipe dimensions and tolerances, note that an outside diameter tolerance of 0.008" either way is allowed on 3" pipe. That's a total of 0.016" from the largest to smallest pipe permitted. In addition to this, an out-of-round tolerance of up to 0.015" is allowed. Note, too, that the pipe's wall thickness may not be less than 0.216", **but it can be quite a bit thicker.**

Pipes skinny. Now suppose that Joe Make-A-Buck Pipe Co. extrudes a truckload of pipes for your local cut-rate distributor. Joe is going to interpret those tolerances to his own benefit. Can't blame him. He's marketing just pipes, not an entire system as Genova does. To use the least materials in producing carloads of pipes, Joe is going to shade the outside diameter down to 3.495", plus 0.005" or minus 0.003", which he can meet. Besides that, he's going to skinny down the pipe's wall thickness to 0.219" plus or minus 0.003".

If Joe doesn't do these things, it could cost him 12 % more to make the pipe. That's more than his profit, so he does this or he goes broke.

Slop factor. Now look at the ASTM fitting dimensions and tolerances portion of Table 4-A. Pete Smith's Fitting Mfg. Co. must follow these, however, Pete has a whole different set of production problems. He knows that his fittings must fit everyone's pipes, even the ones made on poor extrusion machinery, with worse quality control, and

where the amount of material put into a pipe is of secondary importance to getting the pipe out of the machine. So Pete Smith covers himself by staying on the high side of the plus or minus 0.015" fitting tolerances. This way, his 3" fitting socket will fit over all manufacturers' 3" pipes.

If you press a few buttons on your pocket calculator, you'll see the light. A 3" pipe could have an outside diameter as small as 3.492" and still meet specs. A 3" fitting could have a socket entrance diameter as large as 3.535" and meet specs. That's a slop fit of 0.043"! Expressed in fractions, it's half way between 1/32" and 1/16" of slop—far too much for successful solvent welding.

On top of that, if the allowable out-of-roundness in both pipe and fitting happen to fall out of sequence when installed—and they easily could—this adds another whopping slop factor. Then, when a cheap, watery solvent cement is used to solvent weld the joint, it's no wonder that leaks can develop.

Not Genova. On the other hand, Genova doesn't work that way. The Genova team (see page 141) went into the pipe and fittings manufacturing business with the goal of making home plumbing easier and more successful for the home plumber. So they threw away the industry-accepted tolerances. These weren't close enough. Genova holds PVC-DWV pipes to a plus 0.003" minus zero tolerance. And they make their fittings to fit their own pipe, not Joe Make-A-Buck's or anyone else's. That would only end in disaster.

One thing more, you'll notice that most solvent welding instructions tell you: "Test-fit the fitting on the pipe. If it won't go on, or if it goes on loosely and falls off when inverted, you can't make a good joint with it." They're right.

If you use all Genova pipe and fittings, you can forget these instructions. Genova materials fit every time.

So don't ask your dealer whether the yellow pipe will work with the gray fitting. Good DWV systems aren't built that way. Make sure that everything you get is Genova, and be confident that the parts will go together. Enough said.

"Your DWV planning should take into

account the local plumbing codes and what must be done to meet its requirements."

Chapter 5 Roughing the DWV System

5-1 Rough plumbing is exactly that—rough. The below-ground portions of the plumbing for a slab-on-ground house must be put in before the concrete floor is poured. After that, the above-floor parts can be installed.

Rough plumbing is done after the subfloor, roof flashings, and the finished roofing material—shingles—have been installed. Neither insulation nor finished wall, floor, and ceiling materials should be in, however. In a concrete slab structure, all underslab plumbing should have been roughed in before the slab was poured. Rough plumbing begins with the DWV system. Water supply roughing follows, since the smaller tubes can be fitted around the Drain-Waste-Vent piping.

Planning your DWV system is the first step. Toilets are the chief DWV fixtures to plan for, because they need larger 3" soil pipes.

Typical layouts. The five basic plumbing layouts (Fig. 5-2) show compact plumbing setups without additional venting. They will give

you an idea of piping arrangements that can be used. Number 1 is the plumber's ideal fixture arrangement with the washbasin and tub/shower on either side of the toilet. That way, toilet, washbasin, and bathtub/shower can all drain into the vent stack behind the toilet and can also be vented there. Waste pipe runs are short and direct. With the washbasin draining above the floor directly into the vent stack, its liquid wastes wash down past the toilet's drain, helping to keep those pipes open. The washbasin's waste entry uses a reducing sanitary tee (Genova Part No. 61131 in Schedule 30 In-Wall).

Fixture drains and stacks are shown on the plans as white circles, large ones for toilets, smaller ones for the other fixtures. Drains appear as black lines, thick for 3" toilet and

main drains, thinner for 1½" and 2" tub/shower waste branches to the mains. Vent stacks are shown as white circles within partitions.

It's ideal to have another drain-using fixture back up to a bathroom. For example, a kitchen sink, laundry tub, or a second bath or half-bath goes back to back with the bathroom. This makes double use of the first bathroom's DWV and water supply pipes.

In any case, the plumbing should be installed to accommodate the fixtures. It's easy with PVC-DWV pipe and fittings, no matter how much piping is being installed.

Your DWV planning should take into account the local plumbing code and what must be done to meet its requirements. For example, in one national code the maximum for the

5 Basic Layouts

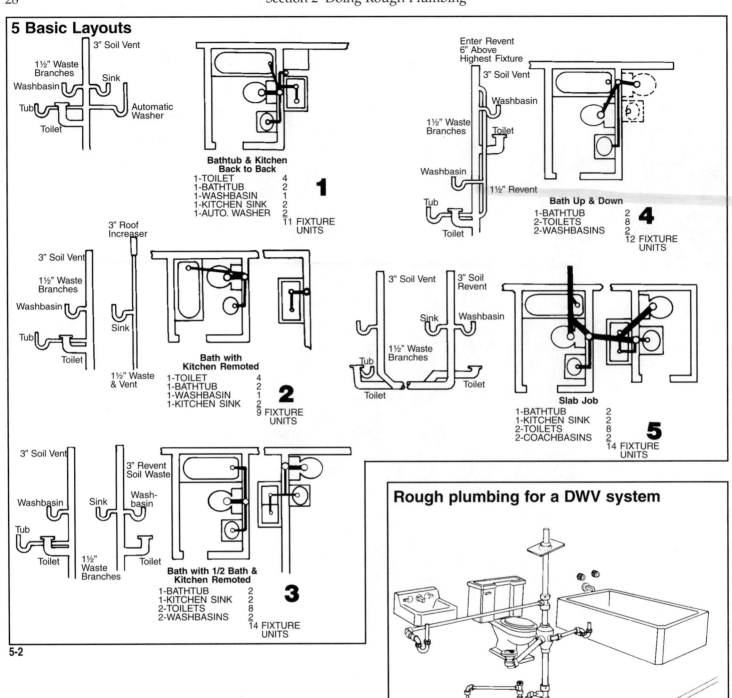

3" Soil Vent

1½" Waste Branches

Washbasin

Sink

Tub

Toilet

Automatic Washer

Bathtub & Kitchen Back to Back

1-TOILET	4
1-BATHTUB	2
1-WASHBASIN	1
1-KITCHEN SINK	2
1-AUTO. WASHER	2
	11 FIXTURE UNITS

1

3" Roof Increaser

3" Soil Vent

1½" Waste Branches

Washbasin

Sink

Tub

Toilet

1½" Waste & Vent

Bath with Kitchen Remoted

1-TOILET	4
1-BATHTUB	2
1-WASHBASIN	1
1-KITCHEN SINK	2
	9 FIXTURE UNITS

2

3" Soil Vent

3" Revent Soil Waste

Washbasin

Sink

Wash-basin

Tub

Toilet

Toilet

1½" Waste Branches

Bath with 1/2 Bath & Kitchen Remoted

1-BATHTUB	2
1-KITCHEN SINK	2
2-TOILETS	8
2-WASHBASINS	2
	14 FIXTURE UNITS

3

Enter Revent 6" Above Highest Fixture

3" Soil Vent

Washbasin

Toilet

Washbasin

Tub

Toilet

1½" Waste Branches

1½" Revent

Bath Up & Down

1-BATHTUB	2
2-TOILETS	8
2-WASHBASINS	2
	12 FIXTURE UNITS

4

3" Soil Vent

3" Soil Revent

Sink

Washbasin

Tub

Toilet

1½" Waste Branches

Toilet

Slab Job

1-BATHTUB	2
1-KITCHEN SINK	2
2-TOILETS	8
2-COACHBASINS	2
	14 FIXTURE UNITS

5

5-2

Rough plumbing for a DWV system

5-3

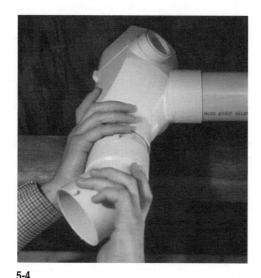

5-4

5-4 DWV rough plumbing starts behind and beneath the toilet. The first part to be installed is Genova's Special Waste & Vent Fitting which snugs between floors to both drain and vent all the bathroom fixtures.

Table 5-A

Maximum Distance Fixture Trap to Vent Stack*

Pipe Size	Distance
1½"	5'
2"	8'
3"	10'

* Check local code

Table 5-B

Design of a DWV System*

Fixture Discharge

Fixture	Fixture Units Units	Trap & Trap Arm Size
Toilet	4	3"
Bathtub/shower	2	1½"
Shower alone	2	2"‡
Sink	2	1½"
Washbasin	1	1¼"
Laundry Tub	2	1½"
Floor Drain	2	1½"

*Check Local Code   ‡ 1½" works well

Table 5-C

Pipe Capacity

Pipe Size	Maximum Fixture Units	
	Horizontal Pipe	Vertical Pipe
1½"	1	2
2"	8	16
3"	35	48
4"	216	256

length of a washbasin's trap arm (trap to vent stack distance) is 5' for the 1½" waste pipe (see Table 5-A). To make this work, the toilet and washbasin have to be close together.

If the bath is on a second story, a Genova Special Waste & Vent Fitting (Part No. 67530 in Schedule 30 In-Wall) can be used in the limited space between floors for hooking up all the bathroom fixtures (Fig. 5-4). This fitting drains as well as vents the fixtures. The SW&V Fitting, which is made only by Genova, enables you to completely customize an installation. Unless a toilet is the **only** fixture to be drained and vented below the floor at that point, we recommend using this fitting. Its branch inlets are 2" socket configurations to accept 2" pipe. Bushings (Part No. 70221) are provided to accommodate the 1½" pipe. There's even a double SW&V fitting made for two bathrooms located back to back. Top caps available for all these fittings contain a single socket for a 3" vent stack or have as many as two additional vents or drains. Any unused side openings

in a SW&V Fitting may be plugged. A plug comes with each fitting.

Pipe-sizing

The first thing to do in planning is to make rough sketches of your house or addition. Very rough is all right, as long as you understand them. A good deal of the plans may even be in your head.

To build a correctly functioning DWV system, you need proper pipe-sizing. Pipes that are too small will not handle the flow; pipes too large will be "lazy," tending to clog. Consult Tables 5-B and 5-C, dealing with fixture units. Fixture units are the accepted measuring stick for plumbing design.

First, size the toilet drain pipe and the waste pipes for each fixture according to these rules of thumb: toilet—3"; shower—2" (some codes permit 1½"); bathtub, kitchen sink, laundry, floor drain, washbasin—1½".

Then size drain pipes that carry the wastes from more than one fixture. Using Table 5-B, add up the fixture units for the groups of fixtures that will empty into a

collective drain. Then see what pipe size in Table 5-C will handle it. For example, if a bathtub, a washbasin, and a kitchen sink all will use the same horizontal drain pipe, that makes a total of five fixture units (from Table 5-B). The horizontal drain pipe necessary to handle five fixture units is 2" (from Table 5-C).

Next, size the main stack. If only two toilets plus other smaller fixtures drain into it, a 3" stack should do easily, but check the tables to be sure. Stacks, of course, are vertical pipes.

Finally, add up the fixture units for all house fixtures and figure how large the main building drain will have to be to carry off the collected wastes. In most cases, a 3" pipe handles it. Building drains are figured as horizontal pipes.

Vents. Size all vents serving toilets at 3". (Some codes permit 2" or smaller toilet vents, but we recommend going up with 3" anyway.) Try to arrange your drainage runs so the washbasin, sink, bathtub, and shower all empty into the main stack above the point where the toilet does. That way, you probably won't need separate vents for these fixtures. Codes call this "stack venting."

In very cold climates, a vent stack that exits through the roof will have to be 3" to prevent ice closure. Smaller stacks can be increased in size 12" below the roof using the appropriate Genova reducing couplings. Top out vent stacks 12" above the roof.

Branch vents. Any fixtures that connect to the drain beneath the toilet's drain must be separately vented. These are called branch vents, or sometimes "revents." Whenever a sink or washbasin is drained below the floor, it must be branch vented. Otherwise, its trap might be siphoned dry by the falling toilet flush water. Branch vent runs may be connected into the handiest vent stacks at least 6" above the flood rim of the highest fixture that drains into that stack. A branch vent pipe should be the same size as the waste pipe for a fixture.

You can run a secondary vent stack up through the roof to vent a fixture. However, it is usually simpler to branch vent it.

Simpler yet is the use of a NovaVent™ Automatic Anti-siphon Valve (see Ch. 14), which replaces an add-on roof vent. Positioned accessibly in place of a vent stack — in the attic or above the flood level of a fixture — they prevent trap siphoning and keep foul air from entering (follow instructions).

Rules

Remembering a few simple rules that plumbers live by will help you to make a smooth flowing DWV system.

Vent-only piping may be connected using short-turn elbows, rather than drainage type ones. These are called vent elbows. Fluid-carrying drain piping **must** use drainage type fittings and drainage piping slopes ¼" per foot in the direction of desired flow. Vent piping slopes upward toward the vent stack the same amount.

Moreover, never use an ordinary tee to connect one horizontal drainage pipe to another. The sharp turn disrupts the flow too drastically. Instead, use a wye placed so that incoming wastes enter in the direction of flow. Fig. 5-5 shows how to make various drainage connections and provide for good flow.

It's a good idea to limit horizontal fixture waste runs (between trap and vent stack) to a total horizontal turn of no more than 90°. Doing this will take some planning.

Possible drain connections

Meeting at an angle — Wye

Parallel — Wye — 45° Elbow

Horizontal Pipes, Same Elevation

Perpendicular — 45° Street Elbow

60° Elbow — 45° Approach — **Better**

Parallel — 90° Street Elbow — 45° Street Elbow — **Good**

45° Elbow — Parallel

90° Street Elbow — Perpendicular

Horizontal pipes, entering pipe should be higher

5-5

Boxing out framing for toilet waste piping

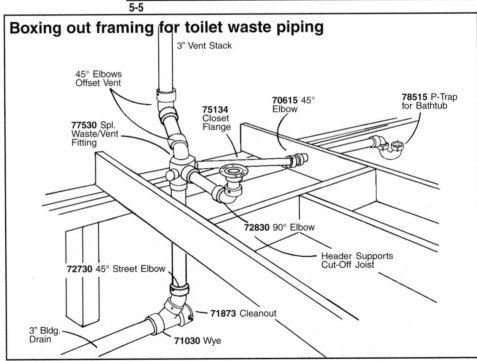

3" Vent Stack
45° Elbows Offset Vent
75134 Closet Flange
70615 45° Elbow
78515 P-Trap for Bathtub
77530 Spl. Waste/Vent Fitting
72830 90° Elbow
Header Supports Cut-Off Joist
72730 45° Street Elbow
71873 Cleanout
3" Bldg. Drain
71030 Wye

5-6

Three Methods of Handling Sink-Washbasin Waste Pipe Above Floor

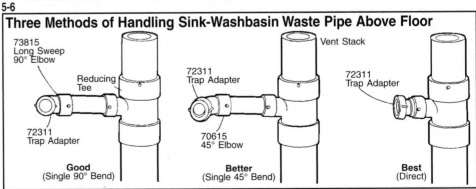

73815 Long Sweep 90° Elbow
Reducing Tee
72311 Trap Adapter
72311 Trap Adapter
Vent Stack
72311 Trap Adapter
72311 Trap Adapter
70615 45° Elbow

Good (Single 90° Bend)   **Better** (Single 45° Bend)   **Best** (Direct)

5-7

Dimensions you need for roughing

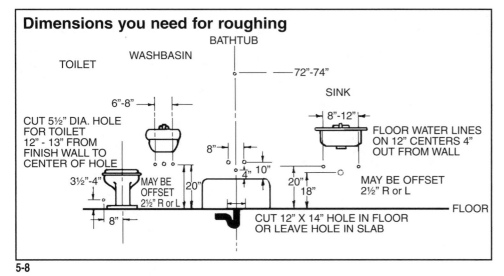

TOILET   WASHBASIN   BATHTUB

72"-74"

SINK

6"-8"

8"-12"

8"

FLOOR WATER LINES
ON 12" CENTERS 4"
OUT FROM WALL

CUT 5½" DIA. HOLE
FOR TOILET
12" - 13" FROM
FINISH WALL TO
CENTER OF HOLE

3½"-4"

4" 10"

MAY BE
OFFSET
2½" R or L

20"

MAY BE OFFSET
2½" R or L

20"

18"

8"

FLOOR

CUT 12" X 14" HOLE IN FLOOR
OR LEAVE HOLE IN SLAB

5-8

Framing a bathroom for fixtures

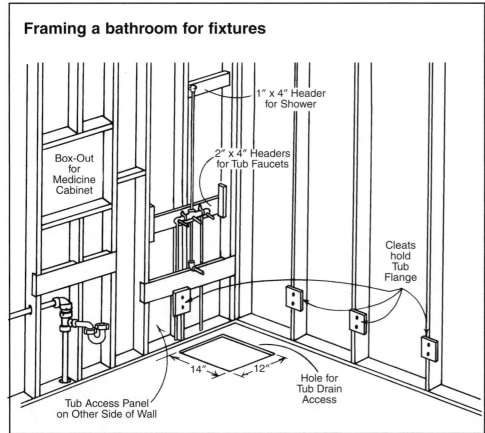

1" x 4" Header
for Shower

2" x 4" Headers
for Tub Faucets

Box-Out
for
Medicine
Cabinet

Cleats
hold
Tub
Flange

14"   12"

Hole for
Tub Drain
Access

Tub Access Panel
on Other Side of Wall

5-9

Rules for Notching Joists   ## Rules for Notching Studs

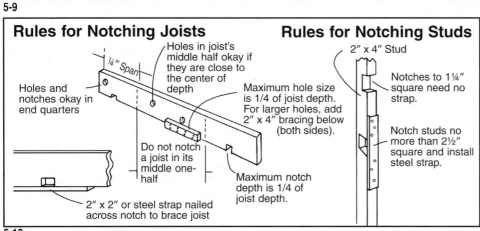

Holes in joist's
middle half okay if
they are close to
the center of
depth

¼ Span

2" x 4" Stud

Holes and
notches okay in
end quarters

Maximum hole size
is 1/4 of joist depth.
For larger holes, add
2" x 4" bracing below
(both sides).

Notches to 1¼"
square need no
strap.

Do not notch
a joist in its
middle one-
half

Notch studs no
more than 2½"
square and install
steel strap.

Maximum notch
depth is 1/4 of
joist depth.

2" x 2" or steel strap nailed
across notch to brace joist

5-10

Fig. 5-7 illustrates three methods. Roughing-in dimensions should be handled as shown in Fig. 5-8. Then you can be sure that all traps and riser tubes will fit them.

Running DWV

The plumbing is half of the project. Fitting it into the house—either going around the framing or through it—is the other half.

As a starter, it's a good idea to mark centerlines for each fixture on the subfloor. Take measurements from the house plan to give you a starting point to see how your piping will fit.

Fig. 5-9 shows the needs of the bathroom fixtures including the wood bracing that should be installed to support each. Regardless of size, PVC-DWV horizontal pipe should be supported at least every 4'. Support vertical runs at every story.

Pipe may be run parallel with framing members by supporting them on headers nailed between members. When running pipes across the framing, the members must be notched. Follow the notching rules shown in Fig. 5-10. Notching should be minimized, as it tends to weaken the floor. Basement and crawlspace cross-member plumbing runs can be installed to the framing faces without notching.

If a floor joist runs right beneath the area that a waste pipe needs to go through the floor, the joist will have to be cut off and a header the same depth as the framing should be installed to support its free end (see Fig. 5-6). If it's practical, the fixture can be moved to miss the joist. Shifting the fixture is easier than installing a header.

Although a toilet's vent stack usually goes up through the wall closely centered behind it, the stack may be located farther away. 10' (developed length) of drain pipe is often the maximum (less by some codes). However, tests have shown that 16' isn't too far with PVC pipe and fittings.

Besides drain pipe space, a bathtub needs an access for connecting its trap. In first floor bathrooms, this can be from a basement or crawlspace below. In second story and slab-type baths, a removable access panel at the head end of the tub needs to be provided during finishing of the walls.

SW&V Fitting. Begin assembling the house DWV system behind the toilet, starting with the Special Waste & Vent (SW&V) Fitting (or a

(continued on page 34)

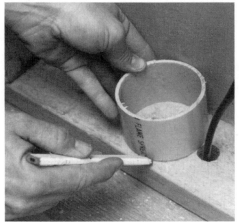

5-11

5-12

5-13

5-11 Plumb the 3" main vent stack behind the toilet using Genova Schedule 30 In-Wall pipes and fittings. No furring of walls is then needed. Marking around a cut-off piece of pipe outlines the hole to be made for the stack.

5-12 If you don't have a proper sized holesaw, an opening for the stack can be made by drilling a series of smaller holes centered on the circle. These should almost touch each other. Be careful not to drill into wiring.

5-13 Then saw out between the holes with a keyhole saw. A chisel will work too. The wood plug will drop out leaving an opening large enough to pass the vent pipe with a little space to spare.

5-14 Use a plumb bob or level to transfer hole locations so your vents run vertically. The vent stack should be close to the toilet.

5-15 Next, cut through the top of the wall and up into the attic. Before drilling, always check to see what's on the other side, as you may not want to bore into it. Wear eye protection when drilling.

5-16 If you're plumbing an add-on room, you'll need access into an existing drain. To get it, saw a piece out of the old drain pipe and install a wye of the same size (see Figs. 5-48, 5-49).

5-17 With a Genova SW&V Fitting or drainage tee beneath the floor, bring up the new vent stack. If the washbasin is close enough, drain it directly into the stack via a reducing tee (Part No. 61131).

5-18 A trap adapter solvent welds to the end of the washbasin's waste pipe. This one required a coupling; another version fits directly over the 1½" waste pipe. The angled waste pipe aims toward the center of the washbasin drain.

5-14

5-15

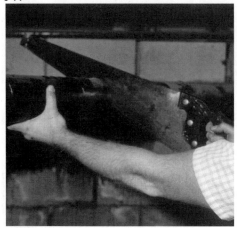

5-16

5-17

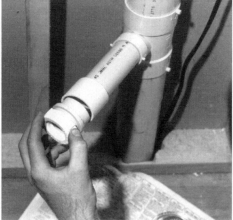

5-18

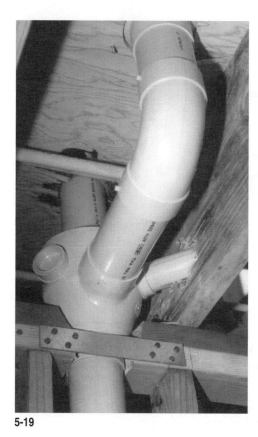

5-19

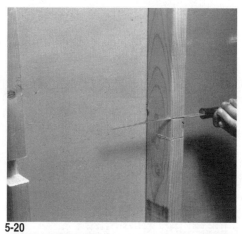

5-20

5-22

5-21

5-19 The amazing Genova Special Waste & Vent Fitting is a problem-solver for DWV plumbing. It accepts the toilet's soil pipe (coming from top of photo), drain (going out bottom), vent (leading up through floor), washbasin and tub/shower drains (entering from the right). While it's designed for use above the first floor, the SW&V Fitting can also be used on the 1st floor.

5-20 Wall framing is notched to pass horizontal drain and vent pipes. To do it, saw in on both sides for the full depth of the notch. Don't go deeper than 2½", though, or you'll overly weaken the wall.

5-21 Then chisel out the wood plug left between saw cuts. Do this by placing the chisel at the bottom of the notch with its wedge toward the notch and tapping with a hammer. If you don't have an all-metal plumber's chisel, you may use a workshop type, being gentle. Wear goggles when chiseling.

5-22 The horizontal pipe can then be solvent welded in place. This one was a drain pipe for a washbasin permitting it to waste above the floor into the 3" vent stack by means of a stack reducing tee.

5-23 Notches should be reinforced and your DWV piping protected from subsequent nailing of wall materials by installing steel straps across them. The 1/16" thick steel straps are available in various lengths.

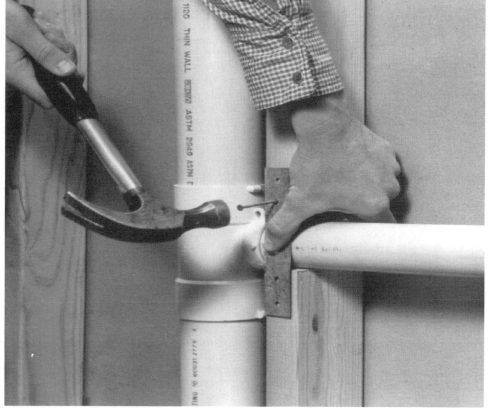

5-23

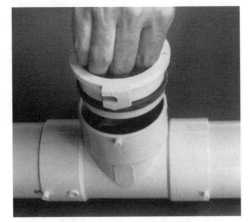

5-24

5-25

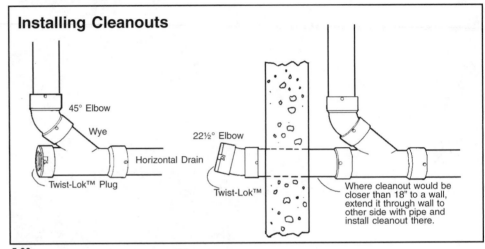

Installing Cleanouts

45° Elbow

Wye

Horizontal Drain

Twist-Lok™ Plug

22½° Elbow

Twist-Lok™

Where cleanout would be closer than 18" to a wall, extend it through wall to other side with pipe and install cleanout there.

5-26

5-24 To use a Genova Twist-Lok™, insert the plug into the fitting hub. Once in, twist it clockwise to lock its two hooks over the projecting pins on the hub. If you wish, you may lubricate the washer with Genova All Purpose Seal Lubricant.

5-25 When plumbing in tight quarters, easy-working PVC piping proves doubly helpful. Run your building drain, installing wyes where needed to let branch drains enter. Be sure to maintain a minimum 1/8" per foot slope as you go (1/4" is better if you have the room).

drainage tee). The exclusive Genova SW&V Fitting is one of my favorites. It'll be yours too, when you experience it. The SW&V Fitting takes the place of a drainage tee under the floor behind the toilet. Made especially for second story bathrooms—but useful on the first floor, as well—the SW&V Fitting accepts a toilet's vent stack and drain at full bore. A pair of side tappings let a washbasin, tub, or shower be drained beneath the floor, while being vented at the same time.

Mark the necessary 3½" hole for the toilet vent run behind it (Fig. 5-11). Also, cut a 5½" hole in the bathroom floor centered about 13" out from the stud wall (12" out from the finished wall) to hold the toilet flange. Now you can run 3" waste and vent pipes in all three directions from the SW&V Fitting or drainage tee. Go up with the vent stack, down with the drain, and horizontally with the toilet soil pipe, as shown in Fig. 5-19.

When running a drain, be sure to install wyes positioned to accommodate any drain entries coming from other stacks. Be sure to install cleanouts at the upper ends of every horizontal drainage run. Above the first floor, cleanouts may be omitted, except for those in a building drain or its branches. Test Tees (71315) are sometimes used in vertical waste

stacks to provide convenient cleanouts in second story applications. Most cleanouts are made using a wye with a Twist-Lok™ plug installed in the open end (Fig. 5-24 & 5-26). Cleanouts must be accessible, and should not be closer than 12" (up to 2") to 18" (larger than 2") to walls they face. Fixture waste pipes don't need cleanouts, as they can be cleaned through the fixture drain or trap.

If you need to offset a drainage run around an obstruction, use two 45° elbows. Offsets in vent-only runs may be made with two 90° elbows. If the offset is slight, and one of the elbows is a street elbow, no short pipe is needed to join them.

In running vent stacks up through walls, install tees at the desired levels for any waste and vent runs that are to enter it. Whenever using a drainage fitting for vent-only use, install it upside down (see Fig. 5-29). This slopes the pipes properly for venting. Genova offers vent (vent-only) tees, so don't get them mixed up with those for drainage.

Carry each vent stack out through the roof with a Genova "Snap-Fit" thermoplastic roof flashing (Figs. 5-36, 5-37).

Fixture wastes. Smaller waste pipes

are installed after the larger 3" pipes have been put in. Work from what's already installed back toward the fixtures. The 1½" washbasin and sink waste pipes extend out through the wall toward the fixture. (See Fig. 5-8 for the rough-in dimensions.) Bathtub waste pipes, of course, go below the floor aimed toward where the tub's drain will go. Install a trap adapter (Part No. 72311) for a bathtub's trap.

Showers are different, as they get 2" waste pipes with solvent-welding rather than tubular traps. A 2" Part No. 78520 (less cleanout) or 78420 (with cleanout) trap is recommended. Get the trap with a cleanout only if it will be accessible. A 2" pipe should run up from the trap through the floor centered on the shower drain opening.

Cut all waste pipes extra long, solvent welding caps onto them to close the DWV system for the water test. Later, before installing the fixtures, these temporary caps will be cut off.

Branch vent runs, if used, are carried up from fixtures via tees. They're elbowed horizontally to enter the vent stack through a tee 6"

(continues on page 37)

Handling Building Drain

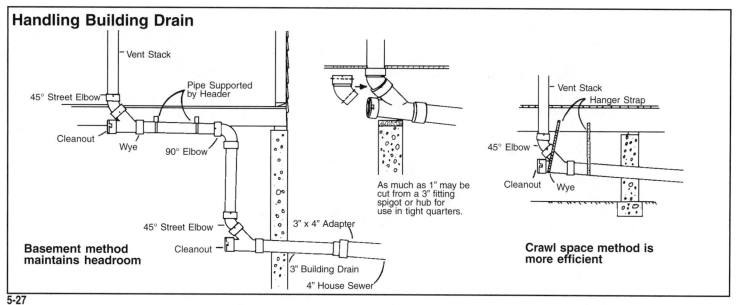

Vent Stack

45° Street Elbow

Pipe Supported by Header

Cleanout

Wye

90° Elbow

45° Street Elbow

Cleanout

3" x 4" Adapter

3" Building Drain

4" House Sewer

Basement method maintains headroom

As much as 1" may be cut from a 3" fitting spigot or hub for use in tight quarters.

Vent Stack

Hanger Strap

45° Elbow

Cleanout

Wye

Crawl space method is more efficient

5-27

Allowance for DWV Pipe Expansion

Less Than 20 Pipe Diameter

Strap Pipe Securely Against End-Wise Movement

Long Run of Pipe

20 Pipe Diameters or more

Strap Pipe Loosely to Permit Movement

5-28

5-29 Branch vent runs entering a stack by a reducing tee may be made with a drainage tee. Intalling it upside down directs gases up and out of the vent.

5-30 Running a vent stack up through the attic and out through the roof goes quickly. If an offset is necessary to miss a roof rafter or other obstruction, you can make it with two elbows and a length of pipe between.

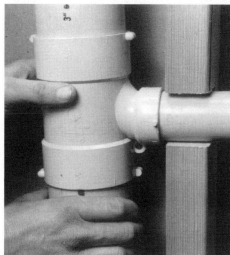

5-29

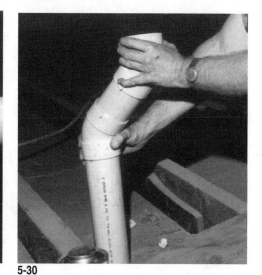

5-30

5-31

5-31 Vent stacks (but not drain pipes) may be offset using pairs of 90° elbows. Offsetting in the attic also can let you bring the vent stack out where you want it on the roof. If you dry assemble your piping in place before solvent welding the joints, you'll know that everything fits.

5-32 The use of a plumb bob ensures that the final vent run out through the roof will be vertical. With the bob centered in the fitting below, the string at the top shows the center of the hole needed in the roof.

5-33 Then if you drive a nail vertically up through the roof sheathing and roofing, the protruding nail atop the roof will tell you where to make the cutout from above.

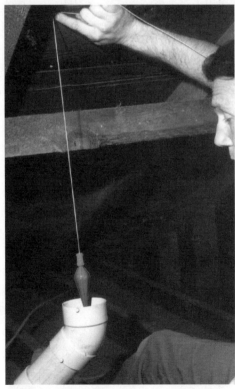

5-32

5-33

5-34

5-35

5-36

5-37

5-34 The cutout can be made from below or above, as you wish. In either case, cut back the roofing in a 7½" (min.) diameter ellipse to clear the Genova "Snap-Fit" flashing base. Saw a 3½" (min.) ellipse through the roof sheathing to pass the vent stack. The ellipse is longer, top to bottom, its length depending on how steeply the roof slopes.

5-35 For topping out, measure a foot above the roof and mark the vent pipe for cutoff. The vent flashing installs under the shingles above the stack and over those below it for leak-free drainage.

5-36 The Genova "Snap-Fit" Roof Flashing is available in 1¼"/1½", 2", 3", and 4" to fit main and secondary vent stacks. The top of the unit's base fits under and over the shingles. Its cap slides down over the vent pipe and snaps onto the base. The flashing won't corrode or stain the roof, and it never needs painting.

5-37 With the cap snapped on over the flashing base, the vent stack is neatly sealed from leaking. No messy mastics are needed.

5-38 Water testing DWV piping tells whether there are leaks. This underground slab-floor drainage system was capped off at all openings and filled with water at its lowest opening. Once tested, the system could be covered with backfill and the house slab poured over it as long as a formal plumbing inspection isn't required. The plugged opening (foreground) is a cleanout for the main building drain.

5-38

above the highest fixture on that stack. Branch vent connections may be made most easily in the attic. Since vent air flow is not restricted, as drain flow is, vent-only runs may include 90° bends in the form of elbows and tees. If it's easier to vent out through the roof, then do it that way. Better yet, if your plumbing code allows, use the appropriate NovaVent™ (Ch. 14).

Testing DWV. When you finish your roughed-in DWV system, you'll want to water test it, for yourself if not for a plumbing inspector. To do this, plug all openings, including the lower end of the building drain. You can make temporary plugs out of mortar stuffed into the pipes against wadded-up newspapers. Let the plugs harden overnight.

Once it's sealed off, you can fill the whole DWV system with water via a garden hose in the lowest roof vent.

If the water level holds and you can find no leaks, call for your rough plumbing inspection. Leave the hose extending up into the lowest vent stack above the roof so that when the inspector comes, he can simply open the sillcock to run some water into the stack. When he sees it immediately overflow, he'll know the system was full and not leaking. The piping plugs can be removed later by chiseling.

Table 5-D

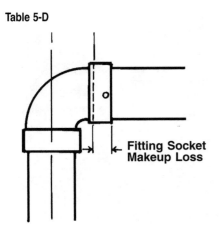

← Fitting Socket Makeup Loss

Makeup Socket Dimensions

Fitting Size	Minimum Socket Depth
1½″ Sch. 40	11/16″
2″ Sch. 40	3/4″
3″ Sch. 30	1½″
3″ Sch. 40	1³/₁₆″
4″ Sch. 40	1⁹/₃₂″

5-41 A No-Hub sleeve connector may be used to join a PVC pipe to a same sized No-Hub cast iron pipe. It is tightened around both pipes sealing the joint watertight.

Face-to-Face Measure

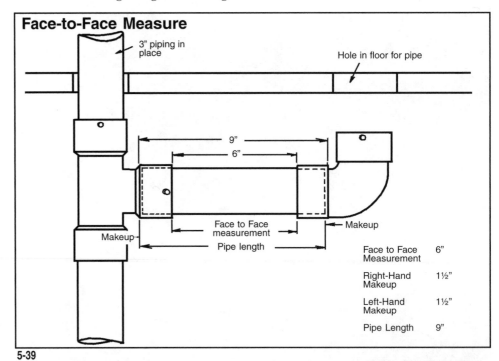

3″ piping in place

Hole in floor for pipe

9″

6″

Face to Face measurement

Makeup →

Makeup

← Makeup

Pipe length

Face to Face Measurement	6″
Right-Hand Makeup	1½″
Left-Hand Makeup	1½″
Pipe Length	9″

5-39

Measuring Makeup

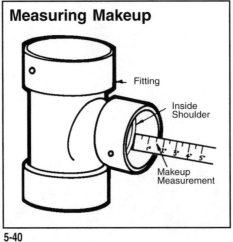

← Fitting

Inside Shoulder

Makeup Measurement

5-40

5-41

Measuring and cutting

Plumbers use several methods for measuring pipe lengths. One is called center-to-center measure and is especially suited to water supply plumbing, and will be covered in the next chapter. In Drain-Waste-Vent plumbing, center-to-center measure is complicated by the wide combination of fitting sizes. For DWV plumbing it's easier to use what's called face-to-face measure.

In face-to-face measure, fittings are held in position on the framing or else laid out on the floor in the same relationship they will have to each other in the framing. Then with a tape measure you measure from the face of each fitting socket to the face of the next one. The pipe must be cut a little longer than that to allow for the distance it slips into the fitting socket at each end. The extra distance is called makeup. Table 5-D

shows the makeup loss for each size of PVC-DWV Genova fittings. Don't forget that actual makeup loss must be doubled because a pipe slides into a fitting socket at both ends.

For example, suppose that a 90° elbow (see Fig. 5-39) is held directly under a hole in the floor where a pipe leading up from it must fit, and its hub is exactly 6″ away from the hub of an already-installed tee. If you cut a pipe 6″ long, there would be no extra pipe to enter both fittings at the ends. Thus the pipe needs to be cut 9″ long with 1½″ of makeup at each end. When finished, the faces of the two fittings will still be 6″ apart, and the upward-facing socket of the 90° elbow will look directly up through the hole in the floor.

If there's any question about the makeup for any fitting, it's easy to measure it right on the fitting socket (Fig. 5-40). But don't forget to add

for makeup or you'll cut pipes too short and have to cut new ones to fit.

Because working with inches and fractions is difficult, you may prefer to use a metric tape measure taking all measurements in millimeters. From this, computations can be made easily on a calculator, as there are no fractions to deal with.

When you cut all your pipes, it's a good idea to dry assemble them with their fittings to see that everything comes out on target. Finally, take your dry assembly apart and solvent weld all the joints.

Adapting

When you need to join Genova PVC-DWV pipe to existing piping of any other type, Genova markets a number of easy-to-use adapters. Here's how to handle various adaptations:

To cast iron pipe. If both the cast iron and PVC pipes are the same

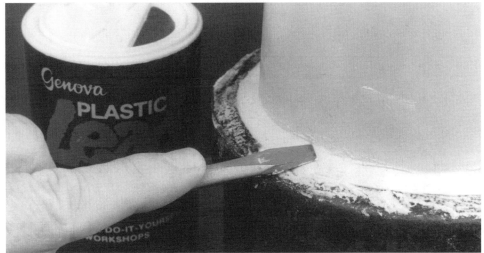

5-42

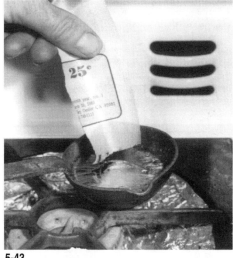

5-43

5-44

5-45

5-42 The best way to finish off a caulked joint is with Genova Plastic Lead. You simply add water and mix. Plastic Lead hardens without shrinking to protect the oakum, holding it in the joint.

5-43 If you caulk with molten plumber's lead, be very sure the lead isn't too hot. To check on this, dip a scrap of newspaper into the melted lead. If it bursts into flame or scorches, the lead is too hot for good caulking. If the paper doesn't discolor, the lead's temperature is just right.

5-44 Quickly pour molten lead into the joint full depth around the plastic pipe. Don't worry. The PVC can take it. But be careful. If there's any water in the joint, the lead will spatter all over you. Be sure to wear eye, face, hand, and body protection.

5-45 Give the joint until the next day to cool. Then you can "caulk" the lead with a tool. Pound it down all around, both against the pipe and against the hub. You'll find it's easier, as well as safer, to use Genova Plastic Lead.

outside diameter, for example 3" No-Hub cast iron hub and 3" Genova Schedule 30 In Wall, use a No-Hub sleeve connector for the coupling (Fig. 5-41). It installs with a socket wrench or screwdriver and even permits slight misalignment between the two pipes.

If the pipes differ in outside diameter, you can get a special flexible rubber coupling (F71915) held with worm-drive band clamps from your local Genova dealer. Each side of the coupling fits one of the pipes.

To cast iron hubs. In each case the pipe should be the same size as the cast iron hub for a good fit. The best way is to make what's called a caulked joint using Genova Plastic Lead (Part No. 14210; for more information on Plastic Lead see chapters 8 and 9.) The use of Plastic Lead saves buying and melting lead. It also keeps you from being burned by hot, molten lead.

Here's how to make a caulked joint: Start by packing rope oakum into the bottom of the joint between the pipe and hub. Tamp tightly using a plumber's "corking" tool and a machinist's hammer. If you don't have such a tool, a large screwdriver may be used as a substitute (wear eye protection). The hard-packed oakum should fill the joint to within 1" of the top. It alone should be tight enough to prevent leaking. The addition of Genova Plastic Lead or molten lead helps keep the oakum in place. The joint should finish level with the cast iron hub. A screwdriver makes a handy applicator. Wipe off any excess Plastic Lead neatly before it sets— in about 20 to 30 minutes. One can of Plastic Lead makes four 3" joints or many more smaller ones. (See Fig. 5-42)

If you use molten lead for covering the oakum packing, make sure it isn't overheated (see test shown in Fig. 5-43). Fill the joint with lead (Fig. 5-44), letting it cool overnight. Then caulk the lead by pounding it down against pipe and hub (Fig. 5-45). Do not caulk while warm or you can force the PVC pipe out of shape.

5-46 Genova flexible adapter (Part No. F71915 in 1½") joins copper DWV to vinyl DWV pipe. Band clamps seal the joints tightly.

5-47 Adaptations to pipe threads can be made with a male MIP adapter. Wrap the adapter's threads with TFE plumber's tape or use pipe dope on them, and thread it in. Tighten with a strap wrench one turn only beyond hand-tight.

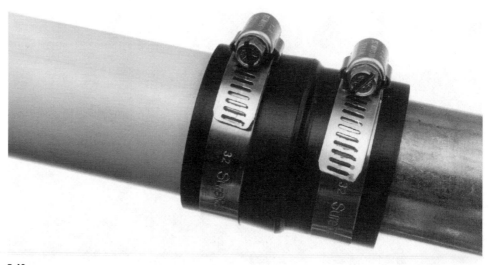

5-46

5-47

While Genova markets special hub adapters in various sizes for use when making caulked joints with molten lead, the PVC pipe itself is amply heat resistant for direct use. This is only if you let the lead cool completely before caulking.

To copper. Genova markets a copper DWV-to-Genova DWV coupling for connecting 1½" PVC-DWV to same sized copper DWV. Part No. F71915 will fit 1½" pipes. For 3" Schedule 40 DWV pipes to 3" copper, use a No-Hub connector. It works as described for No-Hub pipe and Schedule 30 In-Wall.

To ABS plastic. Genova PVC pipes may be attached to existing ABS fittings, provided you do the solvent welding with Genova All Purpose cement. Do not apply too much solvent cement to the ABS, as it may soften too much. Do not use an ABS solvent cement as it has little welding effect on PVC. But remember, most

plumbing codes do not allow mingling different materials such as PVC and ABS.

No ABS manufacturer produces Schedule 30 pipes or fittings, so Genova's 3" Schedule 30 In-Wall pipe needs to be adapted to 3" Schedule 40 before it will fit an ABS hub. To do this, order Genova Part No. 65330 Schedule 30 In-Wall to Schedule 40 bushing. This flush sleeve adapts any Schedule 40 fitting hub to Schedule 30 In-Wall pipe size. In other words, it will make a Schedule 30 In-Wall pipe fit a Schedule 40 socket.

To threads. Genova makes PVC-DWV threaded adapters in male and female threads for all of its pipes. Threaded adapters should be installed with a double wrapping of plumber's TFE tape (or use TFE or regular pipe dope on the male threads). Using TFE tape, wrap the male threads so that threading them in will smooth the tape (Fig. 5-47) on the threads rather

than pull it off.

When used in DWV female threads, a plastic male adapter should be tightened with a strap wrench. This will keep from marring it with the serrated steel jaws of an ordinary pipe wrench. Tighten plastic threaded fittings one turn beyond hand tight. No more.

Warning: Ordinary pipe dope should not be used on ABS threaded adapters— only on PVC. The oils in pipe dope cause stress cracking in ABS, while PVC is resistant to this problem.

To tubular goods. All tubular drainage pipes fit 1½" PVC-DWV pipes with trap adapters. The adapter solvent welds to a 1½" DWV pipe or fitting socket. Use a hub trap adapter (Part No. 72211) to adapt over a pipe; use a fitting trap adapter (Part No. 72311) to adapt into a fitting socket. Both adapters utilize the unique Genova "Dynamic-Duo" washer that

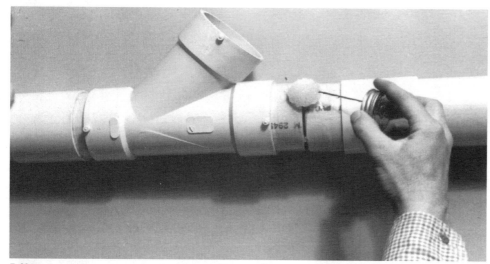

5-48

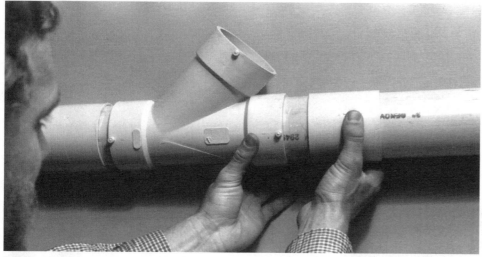

5-49

5-48 Genova offers couplings without inner shoulders. In sizes from 1½" to 4" and 3" Schedule 30 In-Wall, these repair couplings let you join a new fitting, such as this wye branch, into an existing pipe. To do it, slide a repair coupling well back onto each end of the sawed out section of old pipe. Make up the new section to fit, and hold it in place while applying solvent cement liberally on both sides of the joint.

5-49 Then, without delay, slide the repair coupling into place half way astride the joint, giving it a slight twist as you do. Hold the fitting in proper alignment for 10 seconds. That's it.

seals to either 1½" or 1¼" tubular products (see Fig. 11-15).

Genova also offers a cast iron spigot adapter in two types: Hub and No-Hub. As already described, a PVC pipe end works almost as well and is the recommended method because it saves one fitting. But in the case of a 3" Schedule 30 In-Wall pipe being connected to a 4" hub or No-Hub cast iron pipe, you need an adapter that also increases the size. For this, use Genova Part No. 60543 (hub-type) or Part No. 61730 (No-Hub).

To sewer pipes. For connecting 3" DWV pipe to 4" Genova 400 Series sewer and drain pipe, as well as other brands of thin-wall sewer drainage pipes from the house (PVC, ABS, styrene), use a Genova adapter. Get Part No. 61543 for Schedule 30 In-Wall; Part No. 71543 for Schedule 40 and Part No. 71544 connects 4" Schedule 40 to 4" sewer and drain pipe. To connect to vitrified-clay tile, pitch-fiber, and other such pipes, you'll need a flexible rubber coupling.

Existing or new stack? When doing add-on home plumbing, whether you connect your new bathroom fixtures to a main vent that's already there or run a new vent up through the roof, depends on how far away the present one is. We don't advise locating the new toilet any farther than eight feet from the 3" or 4" existing stack, measured along the centerline of the toilet's soil pipes and fittings. In many cases installing a new vent up through the roof is easiest. Or, even better, use the NovaVent™ Relief Valve.

Toilet and washbasin wastes will drain by gravity into the existing building drain. Make the new-to-old drain connection with a wye and slip-couplings (Fig. 5-48, 5-49) or No-Hub couplings (Fig. 5-41).

When it comes to roughing a DWV system, the wide selection of Genova pipes and fittings serves you well.

Compact, trouble-free Universal Fittings

*are so easy to install that once
you try them, you'll be a Genova
plumber for life.*

Chapter 6 Hot/Cold Water Supply Tubes and Fittings

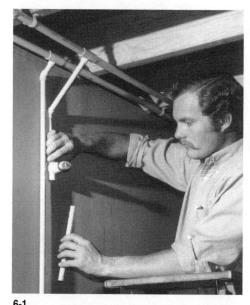

6-1

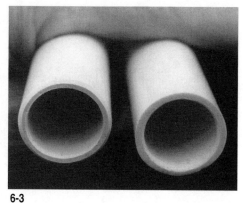

6-3

6-2

6-1 Rigid CPVC comes in 10' lengths, and installs in straight lines for a neat appearance where it will be seen. It's easy-to-install fittings and the tubes are solvent welded into one solid unit.

6-2 Flexible tubing is available in long, fitting-free coils where the tubing can be snaked behind walls, floors, and ceilings. It's also well suited for making a water supply service entrance where it lays in a trench without joints.

6-3 Here's a smooth bore-- not a muzzle loader, but a water supply tube. Genova CPVC is so smooth inside that its flow rate surpasses that of metal piping. Therefore, smaller sizes of CPVC tubing can do the work of larger sizes of steel pipe.

The do-it-yourself way to get new or add-on water supply plumbing for your house is with CPVC (chlorinated polyvinyl chloride) piping. Genova thermoplastic CPVC is vinyl, a rigid heat-toughened version of PVC that serves both hot and cold water. An extra atom of chlorine (Cl-) added to each PVC molecule gives CPVC its heat resistance.

Solvent welding system. Genova offers two systems for water supply piping. Both get you out of sweat soldering. One is the solvent welding system. The other uses Genova Universal Fittings with your choice of tubing. The choice is yours; you will

find working with either system is simple and fast. Facts to help you decide are given in the next chapter on working with the piping.

Genova's solvent welding system features low-cost CPVC fittings (Fig. 6-4a) that join CPVC tubing into one integral system using solvent cement.

The easiest solvent cement to use is Genova's All Purpose Cement.

CPVC can be joined to itself and to copper tubing using Genogrip™ adapters. At one end, these solvent weld into ½" or ¾" CPVC fittings (one Genogrip™ adapter solvent welds over a ½" CPVC tube). At the other end, Genogrips™ adapt mechanically

to CPVC and copper tubing with elastomeric O-rings and metal gripper rings. Both are supplied with the Genogrips™. The O-ring makes a watertight seal, while the gripper ring holds the tube in the fitting under pressure.

Universal Fittings. Universal Fittings incorporate the Genogrip™ adapter concept into elbows, tees, couplings, reducing couplings, adapters, and other useful fittings that let you join CPVC to itself and to copper tubing without solvent welding. These make water supply plumbing as easy as it can be. The fittings are factory-assembled--simply insert the tubing

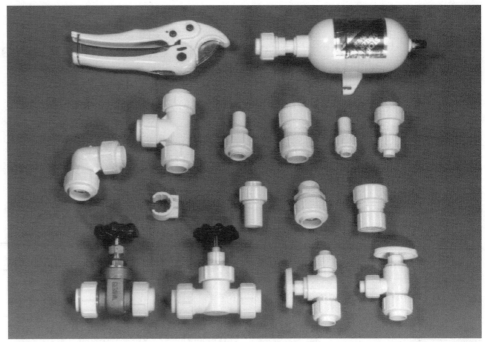

6-4b

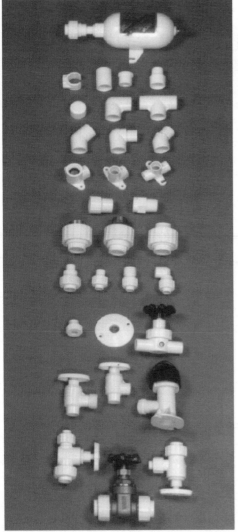

6-4a

6-4a An excellent reason for using Genova solvent-welding hot/cold water supply plumbing products is the vast array of useful fittings available. (l. to r.) (first row) Genova Water Hammer Muffler; (second row) tubing hanger, bushing, cap, coupling; (third row) reducing coupling, 90° elbow, tee; (fourth row) 45° elbow, 90° street elbow, 45° street elbow; (fifth row) winged fittings— shower elbow, elbow, tee; (sixth row) MIP adapter and Special Female Adapter (seventh row) unions— CPVC, MIP transition and FIP transition; (eighth row) Genogrip™ Adapters -- 1/2" and 3/8" street, 3/8", and angled 3/8"; (ninth row) male hose adapter, Special Torque Escutcheon, Color-Coded Universal Line Valve with waste; (tenth row) valves— washer hose valve, angle valve, sillcock; (eleventh row) straight supply valve, 3/4" gate valve, angle supply valve.

6-4b Genova Universal Fittings let you plumb a water supply system without sweat soldering or solvent welding. They are (l. to r.): (top row) Plastic Tubing Cutter, Water Hammer Muffler; (second row) 90° Elbow and tee, Reducing Coupling and Coupling, adapters— Female and Male, and reducing assembly; (third row) 3/4" Gate Valve, Line Valve, Straight Supply Valve, and Angle Supply Valve.

and hand-tighten. They accept directly both CPVC and copper tubing, making CPVC and copper tubing completely compatible without solvent welding or sweat soldering. They easily adapt to threaded pipes and fittings. With Genova Universal Fittings there is never a wait for solvent cement to cure. As soon as the installation is complete, the system is ready for immediate use.

Used as replacements for lead-soldered copper water supply fittings, Genova Universal Fittings also can help you to get the lead out of a copper water supply system.

The easy working, compact Universal Fittings (Fig. 6-4b) are made out of high temperature thermoplastic. Reducing is ingeniously done with a series of assemblies that step the size down to any size needed for the job.

The Genova Universal Fittings system uses the time-tested, trustworthy Genogrip™ design, the Universal system far surpasses the bulkier, less secure fitting systems found on the market today.

Solvent welding CPVC and mechanically coupled Universal Fittings and CPVC tubing make up Genova's unique and complete 500 Series Hot/Cold product line.

CPVC water supply tubing is ASTM rated to **continuously** withstand water at 100 psi pressure and 180° F temperature. It will take periods in excess of 48 hours of 150psi at 210° F.

Genova CPVC is available in 1/2" and 3/4" nominal sizes in 10-foot straight lengths. You can cut CPVC with a plastic tubing cutter (Part No. 534991) or with any fine-tooth saw. Join it to low cost CPVC fittings by easy solvent welding.

CPVC tubes and fittings may be laid below ground in a service entrance trench from a water main or well. Bury the piping (Fig. 7-2) below the frost line. Use solvent welding fittings or Genova Universal Fittings as you prefer.

Genova CPVC tubes and fittings are code-accepted practically everywhere. Some 1 million units in the United States have been plumbed with CPVC. CPVC meets or exceeds the requirements of ASTM D2846. This American Society for Testing and Materials designation specifies tests, methods, and markings for the products. CPVC tubing is approved by the renowned National Sanitation Foundation for conveying potable water, and bears the coveted "NSF" seal. CPVC is accepted by the FHA.

A Genova CPVC water supply system assures a safe, as well as lasting, plumbing system.

Reading a Tee

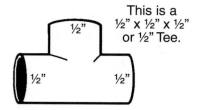

This is a
½" x ½" x ½"
or ½" Tee.

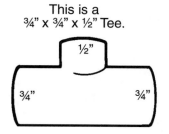

This is a
¾" x ¾" x ½" Tee.

½"

When describing a tee always give the run (through opening) size first, then the branch size.

6-5

6-6

6-6 When you purchase water supply fittings, it pays to be your own quality inspector. Feel inside the fitting sockets for smoothness, rejecting any that are gouged or rough, as these would make for poor solvent welded joints. It's always best to get all your parts by one manufacturer so that you know they will work together.

Called **tubes** instead of pipes, Genova CPVC is sized according to copper water tube sizes. The references given to the various sizes of tubing can be very confusing in that the "nominal" system of identification is used, meaning you have to know how to translate the size given to the tubing's exact dimensions. The only sure way to end up with the right tubing size-- copper or plastic-- is to measure before you buy.

For example, the actual I.D. (interior diameter) of nominal size ½" water tube is ½", while the O.D. (outside diameter) measures exactly 5/8". Similarly, ¾" CPVC tubes scale about ¾" I.D. and measure 7/8" on the O.D. The 3/8" Poly Riser tubing measures 3/8" on the O.D. and ¼" on the I.D., but this is a size that can be confused with another size tubing that is actually ½" O.D. and 3/8" I.D. and is sometimes referred to as "3/8" tubing" and sometimes called " ½" tubing." Again, an actual check of the dimensions you need and specifying both the I.D. and O.D. of the tubing you want is your sure path to success. Tubing also comes in a smaller ¼" O.D. and 1/8" I.D. size for such jobs as hooking up furnace humidifiers and ice makers.

PVC pressure pipe and fittings in Genova's 300 Series are designated by nominal "Iron Pipe Sizes" and thus are not directly interchangeable with nominal "tubing sizes" used in manufacturing CPVC and copper water tubes.

Benefits

Genova CPVC will never corrode, and scale doesn't build on the inner walls, as it does in some kinds of metal piping. Thus, tube bores stay full sized, meaning that an old Genova system enjoys the same fine flow characteristics it had when new (Fig. 6-3). In fact, the Council of

American Building Officials/National Association of Home Builders Plumbing Code indicates that a 3/4" CPVC tube delivers 17 to 37 percent more maximum water flow than a 3/4" copper water tube of the same nominal size.

Moreover, thermoplastic water supply tubing cannot poison drinking water with lead, as sweat-soldered copper tubing has done. In fact, Genova Universal Fittings can be used on copper water tube as lead-free replacements for sweat-soldered fittings. To do it, simply cut out the old fittings and replace them with matching Universal Fittings.

Another benefit: the self-insulating characteristics of Genova CPVC tubes and fittings pay off in preventing sweating of cold water tubes in hot, humid weather. Self-insulating thermoplastic tubing also reduces energy losses in hot water supply lines over metal piping.

Still another advantage, Genova CPVC is a nonconductor of electricity. This makes it free from electrolysis that can destroy a metal water supply system in short order. When dissimilar metals, such as galvanized steel in a water heater and a copper piping system get together, the resulting dielectric action eats away at the whole system. To make things worse, hard water acts like the electrolyte in a car battery to further tear down the metals. None of this can take place with a Genova CPVC system. The nonconducting qualities of Genova materials build into a lifetime water supply system.

When you work with solvent welded CPVC and hand-tighted Universal Fittings, you'll appreciate how they go together without torch welds. Soldering of copper tubes has caused more than one house fire that started by the soldering flame while the water was turned off. Recent U.S.

Table 6-A
Flow through 1/4" ID x 3/8"OD
(Gallons per minute @ 50psi house pressure)

Lengths (ft.)	Flow
2	6.8
3	5.6
4	4.9
5	4.3
6	3.8
7	3.6
8	3.4
9	3.3
10	3.2

Environmental Protection Agency lead-free sweat-soldering requirements are aimed at keeping tin-lead soldered joints from being leached into the potable water system. This requirement makes it that much simpler and easier to plumb with CPVC instead of using soldered copper fittings and tubing. What's more, the other way of getting a water supply system—pipe threading—isn't something you (continued on page 47)

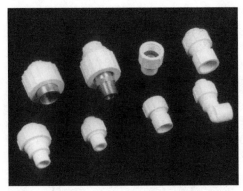

6-7

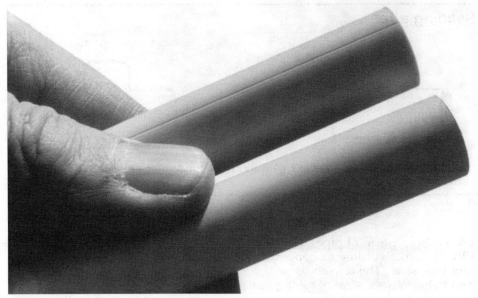

6-8

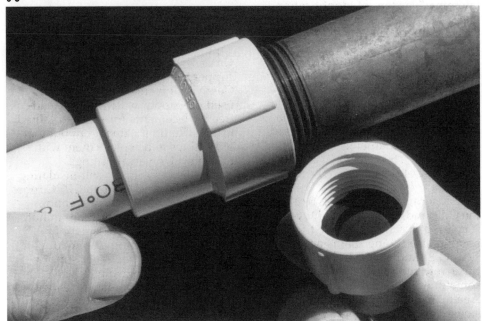

6-9

6-7 Pressurized connections to metal piping and fixtures need transition fittings to absorb differing thermal expansion and contraction without leaking. Shown in the top row (l. to r.) are FIP transition union, MIP transition union, Special Female Adapter, and Universal Female Adapter. In the bottom row are Genogrip Adapters: 1/2″ street CPVC by 5/8″ O.D. , 1/2″ street by 3/8″ O.D., 1/2″ CPVC by 3/8″ O.D., and 1/2″ CPVC by 3/8″ angle adapter.

6-8 Also reject any tubes with visible damage. The upper CPVC tube had a groove running its full length. Making leak free joints with it would be chancy. The Genova CPVC tube (bottom) was typically smooth inside and out.

6-9 The unique Genova Special Female Adapter uses running threads rather than tapered pipe threads. Sealing is provided by the thick elastomeric gasket inside the fitting. Because it squeezes against the pipe for a lasting, leak-free transition, the pipe end should be square and free of scale. Dress it with a file, if need be. No plumber's tape or pipe dope needs to be used on the threads. Just hand tighten.

6-10

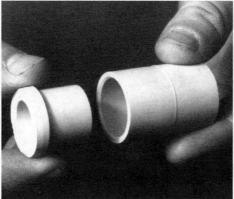

6-11a

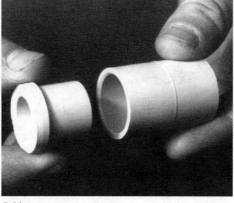

6-11b

Table 6-B

Author's Flow Tests

on Genova's 3/4" x 1/2" Line Valve @50psi

Tested	Flow (gpm)
Line Valve	7.36
1/2" Angle Valve	6.38
3/4" Globe Valve	6.98
3/4" Copper Pipe	7.96

6-10 A plumber's nightmare—tubes that must cross each other on a flat surface—is solved with three 45° CPVC street elbows, plus a regular 45° CPVC elbow. Street fittings fit the sockets of other fittings without the need for short nipples between.

6-11a, 6-11b Two ways of reducing tube size: A CPVC reducing bushing (6-11a) works by taking up space inside the socket of a larger fitting letting a smaller tube be used. A reducing coupling (6-11b) cuts size from 3/4" to 1/2" in a run of tubing. This fitting is also known as a bell reducer.

want to tackle yourself, either.

To top it off, a Genova water supply system costs less. For these reasons, Genova CPVC makes a wise choice for building a first class new or add-on water supply system.

Fittings

One thing you'll like about Genova water supply products is the wide selection of fittings. These join, adapt, reduce, go around bends, divert, turn off, couple, whatever needs doing. (See fig. 6-4a & 6-4b) One thing, make sure that all the fittings and CPVC tubing you buy are of the same brand. This is important, because some dealers stock tubing and fittings manufactured by different firms. You cannot buy a successful plumbing system this way.

If you want a **working system**, don't just get a collection of parts. Specify Genova and be safe.

Reading a tee. The plumbing trade has worked out a simple way of describing tees. It's easy to remember, as well. The sizes of the run sockets—the straight-through portion—are specified first, followed by the size of the tee's branch. For example, a tee having a uniform run of 3/4" and a branch of 1/2" would be called a 3/4" x 3/4" x 1/2" tee (Fig. 6-5, right-hand).

Or a reducing tee with a run size of 3/4" at one end 1/2" at the other, plus a branch size of 1/2" would be a 3/4" x 1/2" x 1/2" tee. A tee with all three openings the same size is called a **regular tee**, or still more simply, a **1/2" tee** (Fig. 6-5, left-hand). This system of reading a tee applies to pressure pipe and Drain-Waste-Vent tees as well as water supply ones.

MIP and FIP. Another trade term you'll encounter is **MIP** for Male Iron Pipe threads and **FIP** for Female Iron Pipe threads. Male threads are those found on the end of a pipe; female threads are those found inside a fitting.

Transition unions. One vital fitting in thermoplastic water supply piping is the hot water transition. One is needed wherever a pressurized CPVC run—hot or cold—connects to metal. The purpose of a transition union is to absorb differential thermal movement between the plastic and the metal without leaking. Simple thread adapters do not take care of this; the transition fitting is different. Genova hot water transition unions (Fig. 6-7, top row, left hand pair) feature two flat faces—a metal one and a CPVC one—clamped together with a threaded CPVC hand-nut. Between the faces, to make the joint watertight, is a thick elastomeric gasket that lets

each face expand and contract while heating and cooling.

Nonpressurized uses, such as the run leading up to a shower head, can be adapted with simple, lower cost MIP adapters. Transition unions aren't needed there.

Special Female Adapter. A unique fitting by Genova is the Special Female Adapter (Fig. 6-9). Also a low cost transition fitting, it solves the thread-leaks problem of FIP plastic threads expanding away from MIP metal threads and leaking, or when being too tight while cool, and cracking. The gasket performs the water sealing chores, with the threads merely keeping pressure on it.

Genogrips. Genova Genogrip Adapters are a unique way to join CPVC and copper tubing with a mechanical coupling. They are transition fittings also; that is, they may be used between metal and plastic. Genogrips come in 3/8", 1/2", and 3/4" O.D. nominal tubing sizes. One end solvent welds to CPVC, while the other joins to CPVC without solvent welding and to copper tubing without sweat soldering.

Genova also makes a number of useful street fittings.

Universal Line Valve. Genova makes its Universal Line Valves with dual-sized 1/2" sockets and 3/4"

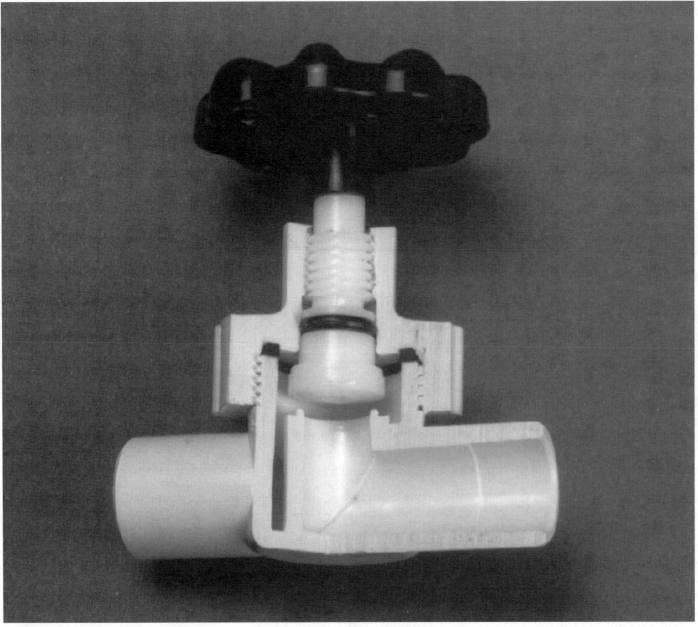

6-12

6-12 Cutaway view shows a Part No. 530151 Universal Line Valve's insides. The unique full-flow valve's passage (coming from the right) runs full bore angled up to the valve seat. Then large passages around the seat lead toward the valve's outlet. Though it is a globe type valve, it gives the flow of a ½" gate valve.

street-sized hubs. These Genova street /socket hubs work with 1/2" or 3/4" CPVC tubes. The hubs fit over a 1/2" tube or into a 3/4" fitting socket. Universal Line Valves fit 1/2" lines direct. The addition of 3/4" couplings at the ends turn them into 3/4" street-sized valves. The line valves may also be used as reducing valves, going from 3/4" to 1/2". The Universal Line Valve is a full-flow valve. (See Fig. 6-12). Table 6-B shows the comparative flow through various items including a Universal Line Valve. These all-around-useful valves come with black handles and with or without waste tappings.

Gate Valve. For locations where full-flow in a 3/4" line is needed, Genova makes a 3/4" gate valve.

This gate valve can adapt directly to 3/4" CPVC or copper tubing by simply inserting the tubing and hand tightening the nuts. Use it in the following locations: service entrance; house main shutoff valve; lawn-sprinkling system; for plumbing a water softener; and for the cold water supply to a water heater (See Fig. 13-28). In other spots, a Universal Line Valve is the way to go.

Wing fittings. Wing fittings are sometimes called "drop-ear" elbows and tees. Their purpose is to provide lugs for rigid fastening of water supply outlets to house framing. They're used as shower arm and bathtub spigot attachments, for automatic washing machine spigots,

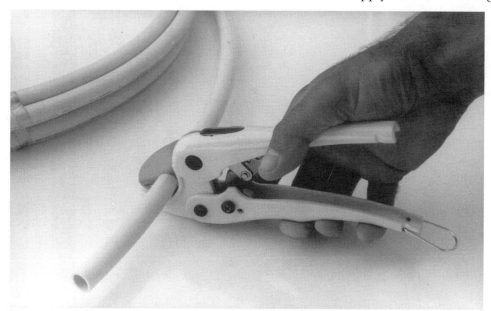

6-13

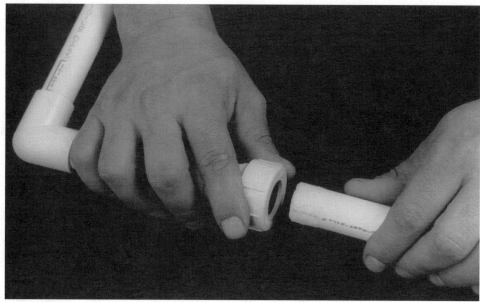

6-14

6-13 To join flexible tubing to its fittings without solvent welding, first cut the tubing. The best way is with Genova's Plastic Tubing Cutter (Part No. 534991) so that there are no burrs to remove. Chamfer the tubing end with a knife.

6-14 Then coat the tube end with Genova All Purpose Seal Lubricant or liquid soap and push it into a Genogrip fitting until it bottoms out on the shoulder. Hand-tighten the nut. You're done.

it is dry, undamaged, and free from grease and oil. If the tube end is dirty, use Genova All Purpose Cleaner. If your plumbing code calls for a colored primer, use Genova Standard Purple Primer (Part No. 13011). Apply a liberal coat of Genova All Purpose or Novaweld® C (required in most code enforced areas) solvent cement to the tube end. Apply a thin coat to the fitting socket, being sure that no bare spots are left. Without waiting, join the tube and fitting full depth, giving a slight twist to distribute the cement. Quickly align the fitting direction and hold for a few seconds. Wipe off any excess cement. It's a good idea to give the final joint two hours to cure before turning on the water.

Mechanical joints

To make up a Genova mechanical connection--either a Genogrip™ adapter or a Genova Universal Fitting-- first see that the tube end is cleanly cut, and free of burrs, and long enough to reach 1¼" into the fitting. It's a good idea to make a mark 1¼" from the end of the tube. Check to be sure the fitting contains all its parts installed in the correct order: O-ring, O-ring retaining collar, and stainless steel gripper ring. Next, coat the tube end with liquid soap or Genova All Purpose Seal Lubricant (Part No. RW206)

Then stab it all the way into the Genogrip™ fitting until it bottoms out on the inner shoulder (Fig. 6-14). Your mark should indicate this. If you find this a hard push, disassemble the fitting and install its parts in proper order over the tube end. Once the fitting is on, tighten the hand-nut and that's it. Genogrips make neat plug-in plumbing connections.

Genova offers a free brochure on CPVC and Universal Fitting Hot/Cold water supply installations. Look for them on the Genova display at your dealer's store. Also available for use are the Genova VHS tapes: "Water Supply Plumbing Basics" and "Plastic Pipefitters Guide." You can order them by calling Genova 1-800-521-7488.

and for washbasin and sink wall stubouts. Genova makes both wing elbows and wing tees in 1/2" CPVC. Wing elbows come in two handy forms, while wing tees come in one.

Genova's Special FIP Wing Elbow (Part No. 530551) is made especially for adapting a chrome shower arm. Running FIP threads—rather than tapered pipe threads—plus an inner elastomeric gasket make for a leakproof seal. The gasket seals tightly against the end of the shower arm.

Ice Maker Kit. For connecting ice makers, furnace humidifiers, reverse osmosis water treatment units, etc. to the water supply, Genova has developed a valve assembly (Part No. 54064). The kit consists of a 1/2" O.D. tee to tap into your water supply line and a 1/4" O.D. valve assembly.

The small, flexible tubing is then routed to the appliance and connected to it. No more need for jury-rigging or brass adapters to hook up small water-using appliances.

Making solvent welded joints

CPVC tubes join to their fittings by solvent welding or mechanical Geno-grip™ fittings, also used in the handy Genova Universal Fittings system.

To solvent weld Genova CPVC water supply joints, first cut the tube to length with a fine-tooth saw or plastic tubing cutter. To remove ridges and burrs from freshly cut tube ends, use a handy Genova chamfering tool (Part No. 534891). It chamfers and removes burrs from the tube end, and will work on both ½" x ¾" CPVC tubing. Check the tube end to be sure

Take time to rough the water

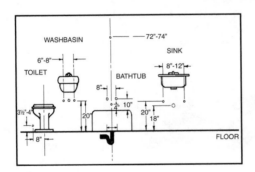

system on paper room-by-room before you begin.

Chapter 7 Roughing the Water Supply--Room by Room

Table 7-A

Water Supply Sizes for Fixtures
Poly Riser

Fixture	Service	Tube Size	Supply Tubes Size (O.D.)
Toilet	cold	1/2″	3/8″
Washbasin	hot/cold	1/2″	3/8″
Kitchen Sink	hot/cold	1/2″	3/8″
Laundry Tub	hot/cold	1/2″	3/8″
Bathtub	hot/cold	1/2″	
Shower	hot/cold	1/2″	
Dishwasher	hot	1/2″	3/8″
Automatic Washer	hot/cold	1/2″	
Sillcock	cold	1/2″	
Service Entrance	cold	3/4″	
Hot/Cold Water Main	hot/cold	3/4″	
Water Heater	hot/cold	3/4″	
Water Softener	cold	3/4″	

Running the tubes and fittings for a CPVC hot/cold water supply system is easy. So easy that, until now, no one book on plumbing has bothered to put together complete pipe fitting information on it. In this book, we do that for you. Here's what you need to know on flow, fittings, measuring, plus installation requirements to use Genova's popular solvent welding and Universal Fittings used with rigid or flexible water supply piping.

Tube sizing. Water rolls down the slick walled Genova thermoplastic tubing so readily that you can forget about assigning fixture demand units, adding up pipe fitting flow resistances, and consulting complex flow charts for house water supply plumbing. This is left for other piping materials. These Genova tubes flow so well that a set of simple rule-of-thumb tube sizes will let you design and build a hassle-free house plumbing system without flow problems.

For single family residence use, see Table 7-A for the tube sizes to use. It takes into account the fact that larger water-using fixtures and appliances need larger supply piping than lesser ones.

In general, here's how to size your system. Use 3/4″ I.D. tubing from the service entrance through the water meter, main shutoff valve, and water softener, if any (some codes call for a 1″ service entrance, but, with 40 psi pressure or more, plus the use of water saving fixtures, that seems excessive). Go on to the water heater. Call this the cold water main. From the water heater, a hot water main begins, paralleling the cold water main. Both should be in 3/4″ tubing. Branch off of these to the fixtures in 1/2″. Reducing tees in CPVC (3/4″ x 3/4″ x 1/2″—Genova Part No. 51471) do the job. In Universal Fittings use Part No. 545071 tees with Part No. 540011 reducing assemblies in the tee branches. At the next to last branch off, a 3/4″ x 1/2″ x 1/2″ reducing tee (Part No. 51473) or

Universal Tee with two reducing assemblies will terminate the main into a final pair of 1/2″ branches. Make each branch run untapped all the way to its fixture drop.

To keep garden hose use from stealing too much water pressure from house fixtures, make hose outlet runs in 1/2″. Take them from the cold water main as close as you can to the water heater. Don't use any end-of-main hose takeoffs, and never take them from the 1/2″ branches serving fixtures. If a run serves two sillcocks, you can make it from 3/4″ tubing, but reduce to 1/2″ where the branches separate. Of course, if you soften your water, all sillcock runs should exit before the softener (you may want to install an unsoftened drinking water tap in the house, especially if a family member's intake of sodium must be restricted).

Genova makes a unique sillcock: Part No. 530941. It solvent welds directly to 3/4″ or 1/2″ CPVC tubing

Water distribution system using CPVC

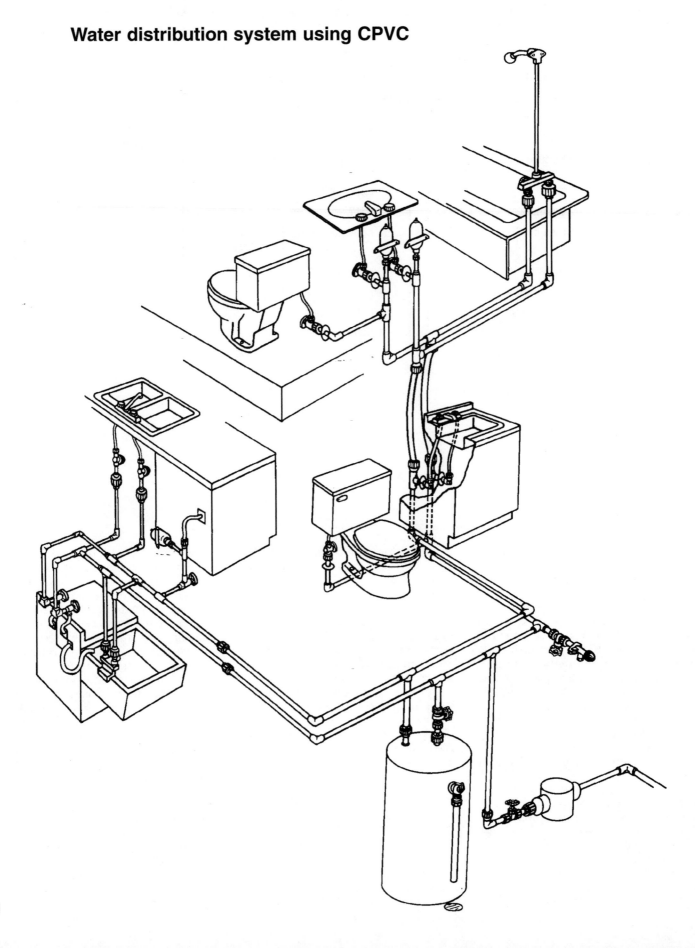

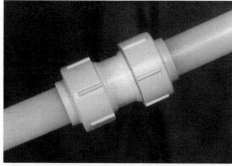

7-2

7-3

7-4

7-5

with a bushing, and has flanges for firm attachment to the house wall.

Should house water pressure be more than 80 psi, install a pressure regulator set to 50 psi, the ideal pressure. It goes in right after the water meter. Outdoor water taps may come before or after the pressure regulator, depending on how much pressure you want them to have. Outdoor water in sprinklers is generally easier to handle and less wasteful at 50 psi.

Making a Plan

Fig. 7-1 shows a water distribution system in 3/4" and 1/2" CPVC. Naturally, the actual runs will depend on the location of the water service entrance and water meter, as well as the fixtures and appliances in your house. Moreover, the piping arrangement will depend on whether the house is single or multi-story and whether the plumbing comes from a basement or crawlspace, or from an attic.

Plan your system in the direction of water flow, locating the fixtures and then the pipes necessary to serve them. Space hot and cold lines about 6" apart. Never cross connect the water supply system to the drain-waste-vent system or to any other source of contamination, such as a

lawn sprinkling system.

If you use isometric paper to lay out your system, each pipe direction can be plotted along a different set of lines, showing the whole system clearly. But you don't need to be that organized. The main thing is to have in mind what you plan to do. I recommend simply sketching out the system, not necessarily to scale. Show all fixtures and appliances, all pipes and fittings, labeling them. Pipes can be just single lines.

7-2 Flexible tubing water service entrance can be installed fitting-free below ground. When laying it in a below-frostline trench, snake the tube back and forth a little to allow extra tubing for contraction when cold.

7-3 For a service entrance run of more than 100', lengths of flexible tubing can be joined in the trench with a Part No. 541071 Universal Coupling.

7-4 For water-testing a supply system before the water hookup is made, a hose adapter made from a Genova Part No. 543071 Universal female adapter is handy. That way the water supply system can be pressurized with a garden hose. It connects directly to the end of a garden hose with two or three garden hose washers used in the adapter.

7-5 It's best if water supply tubes that pass through a concrete slab are kept from bonding with the concrete. Here a wrapping of bond breaking plastic electrical tape is made.

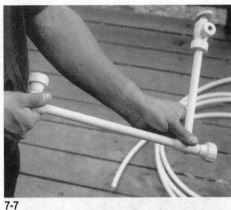

7-6 A service entrance tube may be covered by concrete, but it should run below the concrete, not within it. Risers coming to the top of the slab can be staked to hold them in position. When the above-slab plumbing is done, it can be joined to the risers with a coupling.

7-6

7-7

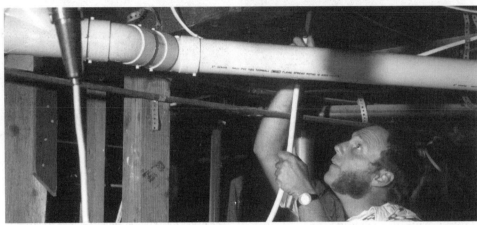

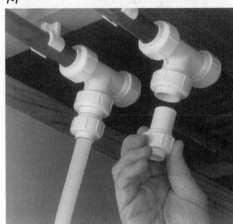

7-8

7-9

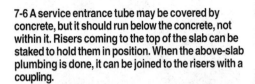

Tapping into a copper main with a Genova Universal Tee

existing ¾" copper main

¾" tee 545071

remove 1½" of copper main

¾" copper pipe

copper ¾" tee

¾" copper pipe

copper ¾" elbow

Reducing assembly used to reduce from ¾" to ½" tubing 540011

½" CPVC tubing

Tapping into a galvanized steel main with CPVC

saw and remove threaded pipe section

galvanized tee

¾" CPVC pipe

¾" CPVC pipe

galvanized elbow

existing ¾" galvanized main

MIP transition union 530501

¾" x ¾" x ½" reducing tee 51471

Coupling 50105

CPVC ½" tubing

Genogrip™ Adapter 530751

MIP Transition Union 530501

7-10

7-7 Solvent welding of CPVC subassemblies ahead of time makes the job go easier. Here, a hot water branch pipe is being put together with Genogrip Adapters to accept longer runs leading to and from the CPVC branch.

7-8 Making add-on hot and cold water mains with 3/4" CPVC minimizes the time you need to spend working in a crawlspace. CPVC is flexible enough that it can be threaded behind walls, floors, and ceilings.

7-9 For new hot and cold water runs, tap into existing copper mains with pairs of Part No. 545071 3/4" Genova Universal Tees that fit directly to the copper mains. The tee branches each accept a 3/4" or a 1/2" branch if a reducing assembly is used.

Proper design

Keep in mind a few rules for designing your water supply system as you go.

Support tubes on 32" maximum centers (every other joist).

CPVC tubes should not be restrained against thermal expansion and contraction. A 10' length will expand about 1/2" when hot water flows through it. To provide for that movement, do the following:

1. Always use Genova tubing hangers or straps to support your system (see Fig. 7-11a, b, c). These hold the pipe snugly to the framing, yet permit end-wise movement. They also will not cut into the pipe. Use one support at every other joist, or 32 inches on center.

2. Never install long runs of CPVC tube that are bound in by walls or framing. Leave a little space at the ends of runs for expansion (Fig. 7-13).

3. Make 1' doglegs (offsets) in overly long runs of CPVC pipe, those more than 35' long. These will bend enough to take up expansion (see Fig. 7-12).

4. CPVC risers branching off of CPVC mains should be unbound and long enough from the main to where they extend through the floor or ceiling to accommodate slight movement of the main. An 8" distance should do the trick.

5. All piping should be installed in heated spaces, or else heating cables should be provided to keep water in it from freezing. If the water in a CPVC tube should freeze, the best way to thaw it is with hot water poured over cloths wrapped around the piping. No matter how thick, insulating of piping will not prevent freeze ups.

To allow movement, holes drilled for water supply tubes need to be larger than the tubes: use a 7/8" bit for 1/2" tube and a 1" bit for 3/4" tube. High-speed wood-boring bits chucked in an electric drill do the job easily.

Wherever CPVC pipes come within a nail's length (1¼") of the face of a framing member, or where the framing is notched out for them, install 1/16" thick metal straps (Fig. 7-16). These will protect the tubes from nail penetration.

The use of Genova Water Hammer Mufflers (Part No. 530901) at fixtures and water-using appliances is a must. Codes require such protection against water hammer. Chapter 14 tells how to install them.

Don't forget to use transition fittings at all pressurized hot and cold water connections to metal. Genova Transition Unions (Figs. 7-17 to 7-20), Genova Special Female Adapters, and Universal Female (but not male) Adapters may be installed directly to a water heater's fittings. Other brands of transition fittings need 8" to 11" long metal extension nipples on the heater to dissipate burner heat before it can reach the fittings. Whenever using any threaded metal pipe in a CPVC system, it's best to go for costlier brass. Using galvanized steel might limit the life of your system, as it can corrode.

Be sure to make your vertical supply runs to fixtures—which are called **fixture drops**—with the hot water pipe on the left as you face the fixture and the cold on the right. Install Genova Gate Valves (Part No. 530271) as a main shutoff after the water meter and as a cold water supply shutoff at the water heater. Also, install a valve at every fixture on both the hot and cold water sides. Use either Part No. 530651 angle supply

(continued on page 58)

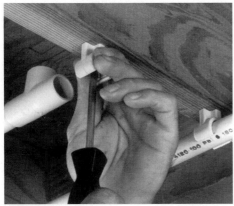

7-11a

7-11b

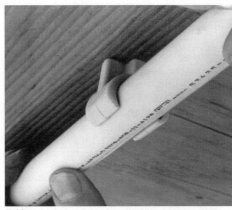

7-11c

7-12

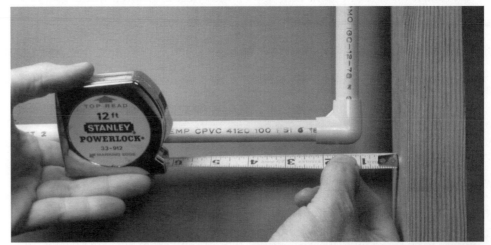

7-13

7-11a,b,c One of a kind Genova tubing hangers incorporate a special design that allows tube expansion. Install with #6 x 1″ drywall screws, which need no pilot holes (7-11a). Snap the tube into the installed hangers (7-11b). Then the tube will be firmly supported, yet free to move with expansion and contraction (7-11c).

7-12 To allow for expansion and contraction of long runs of tubing, make foot-long doglegs in runs more than 35′ long.

7-13 And don't bind tubes in. Allow an inch of space between tubing runs and house framing for expansion.

7-14 Tubes going upward from mains, such as this valved one, are called risers. (In an attic installation, they go downward.) The risers should run for at least 8″ before passing through a hole. Otherwise, they will restrict thermal movements of the main.

7-15 Holes for water supply tubes are easily drilled using a high-speed wood-boring bit in an electric drill. Make the holes 7/8″ for 1/2″ tubes, 1″ for 3/4″ tubes. This allows room for thermal movements of the piping.

7-14

7-15

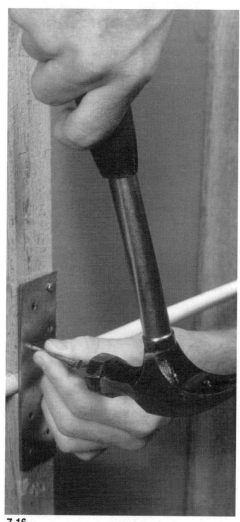

7-16

7-17

7-18

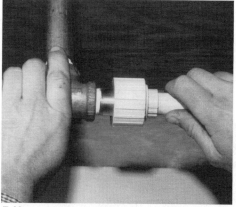

7-19

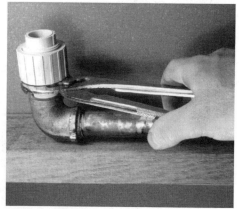

7-20

Table 7-B
Bending of CPVC Tube

Size	Min. Bending Radius
3/8"	8"
1/2"	18"
3/4"	18"

7-16 Water supply lines need protection from the nails used to hold wall and ceiling materials. Notches in house framing made for water supply tubes also need reinforcing. Both jobs are accomplished at once by nailing 6"-long 1/16"-thick punched-steel straps over the framing at all notches.

7-17 To adapt CPVC to threaded metal pipes, simply install the proper sized hot water transition union on the threaded side. Tighten with a pipe wrench. TFE tape or pipe dope helps threaded joints to seal tightly and also helps them to come apart easier, if that's ever necessary. If the transition union will be inaccessible later, give the hand nut an extra quarter "insurance" turn with water-pump pliers.

7-18 With the hot water transition union installed, start putting in CPVC tubing, solvent welding instead of threading the joints. Apply solvent cement to the tube end, then the fitting socket.

7-19 Without waiting, push the tube end all the way into the fitting to complete the solvent welded joint. You're now started with CPVC, and can continue using it to the end of the run. No more adapters are needed.

7-20 Where working quarters are close, transition unions can be tightened with water pump pliers. In any case, don't neglect to use a transition fitting when connecting pressurized CPVC hot and cold water piping to metal.

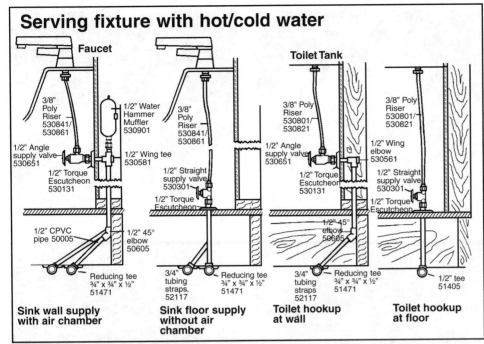

Serving fixture with hot/cold water

Faucet

3/8" Poly Riser 530841/ 530861

1/2" Water Hammer Muffler 530901

1/2" Angle supply valve 530651

1/2" Torque Escutcheon 530131

1/2" CPVC pipe 50005

1/2" 45° elbow 50605

Reducing tee ¾" x ¾" x ½" 51471

Sink wall supply with air chamber

3/8" Poly Riser 530841/ 530861

1/2" Wing tee 530581

1/2" Straight supply valve 530301

1/2" Torque Escutcheon

3/4" tubing straps 52117

Reducing tee ¾" x ¾" x ½" 51471

Sink floor supply without air chamber

Toilet Tank

3/8" Poly Riser 530801/ 530821

1/2" Angle supply valve 530651

1/2" Torque Escutcheon 530131

1/2" 45° elbow 50605

3/4" tubing straps 52117

Reducing tee ¾" x ¾" x ½" 51471

Toilet hookup at wall

3/8" Poly Riser 530801/ 530821

1/2" Wing elbow 530561

1/2" Straight supply valve 530301

1/2" Torque Escutcheon

1/2" tee 51405

Toilet hookup at floor

7-21

7-22

7-22 When plumbing for a one-piece toilet, the water supply should come out as low on the wall as possible. It merely needs to be high enough off the floor so that a Genova Torque Escutcheon (installed during finish plumbing) will clear the base trim. A good way to make this installation is with a wing elbow screwed to a 2" x 4" header resting on the wall's sill plate. It holds the CPVC stubout securely.

valves, or Part No. 530301 straight supply valves. For a sink supplied in Universal Fittings, a Part No. 540051 Line Valve is often used. Use valves on entering any type of water-using appliance, like a water softener. This is so you can turn off the water when taking the appliance out for service or replacing it.

As part of your water supply installation, fit horizontal 1" x 4" boards into the notched studs, nailing them solidly behind washbasin, sink, bathtub, and shower. Fig. 5-9 shows what's generally needed. Get each at the correct height to do its job. These, if not used to support a heavy fixture, will brace the water supply piping. It's best, though not necessary, to have the plumbing fixtures on hand when you do the rough plumbing.

Fig. 7-21 shows four methods of serving fixtures with hot/cold water. Figs. 7-22 and 7-23 show how to handle a one piece toilet's water supply.

Rough-in measurements

Perhaps the best way to run water supply pipes is to work in the direction of water flow. As you branch off to fixtures, you can either go right on through building the mains or you can complete the branches, then continue with the mains.

Fixture drops. When you reach a fixture, create what are called **fixture drops**. These include the Water Hammer Muffler (not needed at toilets) and—in a solvent welding system—the wing elbow or wing tee that sends a stubout pipe through the wall to supply the fixture with hot or cold water. We recommend using solvent welding fittings for making fixture drops because they provide the rigidity of wing fitting attachments.The locations of through-the-wall stubouts to fixtures is called the water supply rough-in. They vary depending on the fixture. Water supply rough-ins are shown in Fig. 5-8.

Washbasin rough-ins. You don't want your hot and cold water rough-ins too close together. A 6" separation is workable, 8" is better. If you make them 20" high, they'll be at the same level as the waste opening. In a solvent welding system, use Genova CPVC wing tees combined with Genova Torque Escutcheons to make a good working, firmly attached piping installation.

Toilet rough-in. Before you can rough-in a toilet's water supply, you need to have the toilet on hand. Toilets vary widely in how high a rough-in they need. Generally, the lower the better—2" to 2½" above the floor. Fig. 14-41 illustrates a good, low solvent welded rough-in where the pipes are supported on a 2" x 4" header nailed to the plate.

Shower arm rough-in. Install a 2" x 4" header between two studs for support. It's best to do a shower's mixed water riser in solvent welding CPVC. Rough in the shower's Special FIP Wing Elbow (Part No. 530551) 72" to 74" above the finished floor.

Piping measurements

In a CPVC system, piping measurements may be taken in either (or both) of two ways: center-to-center or face-to-face. For water supply piping, center-to-center measure has the most advantages. With it, each fitting produces a gain in distance. Fitting gains are simply subtracted before cutting the tubes.

Gains varies. It may be measured on each fitting or taken from Table 7-C, which gives the gain in both inches and millimeters. If you have a metric ruler, you may prefer working that way. It gets you out of dealing with hard to handle fractions. Metrics can be manipulated on a pocket calculator.

Face-to-face measure has the advantage of your having to remember only two figures: add 1" to 1/2" tube for makeup, the distance the tube slips into the fitting sockets, and add 1³/₈" to 3/4" tube. These amounts cover the makeup at both ends of the tube. So, with it, you simply measure face to face between fittings, add the makeup, and cut the tube to that length. Table 7-C also shows the makeup for various fitting sockets.

For a comparison of the two measuring methods, see Fig. 7-24. With center-to-center measure, if you forget to deduct for fitting gain, the resulting too-long tube can be cut again and used. But with face-to-face measure, if you forget makeup, the

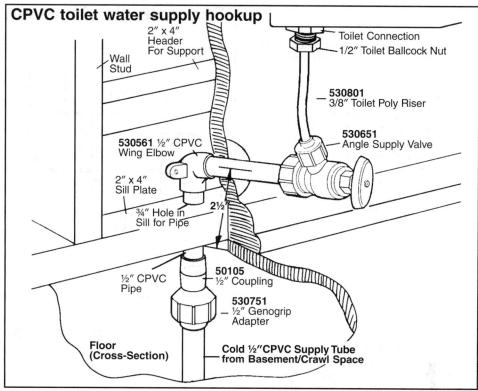

CPVC toilet water supply hookup

- 2" x 4" Header For Support
- Wall Stud
- Toilet Connection
- 1/2" Toilet Ballcock Nut
- 530801 3/8" Toilet Poly Riser
- 530561 ½" CPVC Wing Elbow
- 530651 Angle Supply Valve
- 2" x 4" Sill Plate
- ¾" Hole in Sill for Pipe
- 2½"
- ½" CPVC Pipe
- 50105 ½" Coupling
- 530751 ½" Genogrip Adapter
- Floor (Cross-Section)
- Cold ½"CPVC Supply Tube from Basement/Crawl Space

7-23

Table 7-C
CPVC fitting measurement

Fitting	Fitting gain in. (mm) (center to tube end)		Makeup-- in. (mm) (socket depth)	
	1/2"	**3/4"**	**1/2"**	**3/4"**
90° elbow, tee	3/8 (10)	9/16 (15)	1/2 (12)	11/16 (17)
45° elbow	5/16 (8)	3/8 (10)	1/2 (12)	11/16 (17)
90° street elbow (street side only)	1 (25)	1 3/8 (35)	1/2 (12)	11/16 (17)
Coupling	1/8 (3)	1/8 (3)	1/2 (12)	11/16 (17)
Universal Line Valve	1 (25)	13/16 (20)	1/2 (12)	11/16 (17)
Union	5/8 (16)	9/16 (15)	1/2 (12)	11/16 (17)
¾"x¾"x½" Reducing Tee	3/4 (19)	1/2 (12)	1/2 (12)	11/16 (17)
¾"x½"x½" Reducing Tee	1/2 (12)	1/2 (12)	1/2 (12)	11/16 (17)

tube will have been cut too short to fit
and you'll have to start with a fresh
piece.

Pipe offsets

Figuring of piping offsets where 90°
and 45° water supply fittings are
involved is easy. For 90° fittings take
direct measurements. For 45° offsets
consult Fig. 7-25. Multiply either the
set or the rise (with 45° elbows
they're equal) by **1.414**. Don't forget
to deduct for fitting gain before
cutting the tube.

To make neat appearing, parallel 45°
offsets in rigid CPVC hot and cold
water mains (Fig. 7-26, left-hand), you
need to know how much longer to
make the outside tubes than the
inside ones. To find distance B, simply
multiply the distance between the
tubes (distance A) by 0.414. For
example, if the tubes are 6" apart, the
outside tube would be cut (6 x 0.414)
2.484" longer than the inside one
(2½" longer should be close enough).
Then outer tube D is cut longer than
inner tube C by 0.828 x A inches. For
example, with the 6" tube separation,
if inside tube C measures 4", outside
tube D would be [(0.828 x 6)+4]
8.968" long. Fitting gain is not a factor
except in positioning the first fitting.

To make parallel offsets with 45°
elbows (Fig. 7-26, right hand), first
figure B, as described above. Then,
cutting tubes E and F the same length
creates the rest of the offset
automatically. There's nothing to it.
Distance B, already calculated for the
incoming tubes, remains the same for
the outgoing ones but with the
position of the longer tube switching
sides. Again, fitting gain plays no part.

If calculations become confusing,
just lay the fittings out on the floor
the way you'll want them and take
measurements.

Handling details

As with much pipe fitting, before
solvent welding them, it's a good idea
to dry assemble the parts to see that
everything fits. Later you can solvent
weld each joint. Don't forget any
joints, or you'll have a leak. If you
make a mistake and solvent weld a
wrong fitting, all you need to do is
saw out the error, replacing it with the
correct fitting and two couplings. The
couplings will join the new to the old.
How to do this is shown in Fig. 7-27.

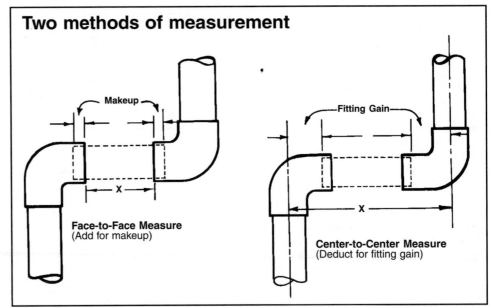

Two methods of measurement

Face-to-Face Measure
(Add for makeup)

Center-to-Center Measure
(Deduct for fitting gain)

7-24

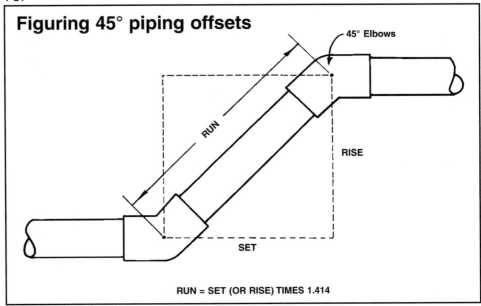

Figuring 45° piping offsets

45° Elbows

RUN

RISE

SET

RUN = SET (OR RISE) TIMES 1.414

7-25

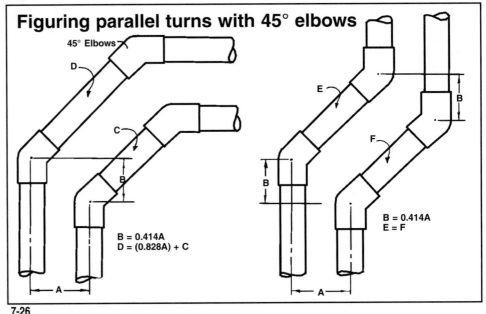

Figuring parallel turns with 45° elbows

45° Elbows

D

C

B

B = 0.414A
D = (0.828A) + C

A

E

B

F

B = 0.414A
E = F

B

A

7-26

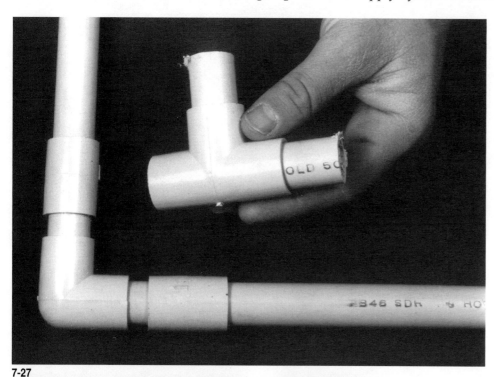

7-27

7-27 If you goof and install the wrong CPVC fitting or, by failing to follow good solvent welding practice, get a leaky joint, it's easy to fix. Simply saw out the wrong section of tube, installing a new section with couplings to join it in. That's all there is to it.

7-28 At tub/shower mixing valves, as well as individual valves, water supply tubes are fitted via MIP transition unions. TFE tape or pipe dope is used on the male threads. Note how the 2″ x 4″ support header has been notched to clear the unions.

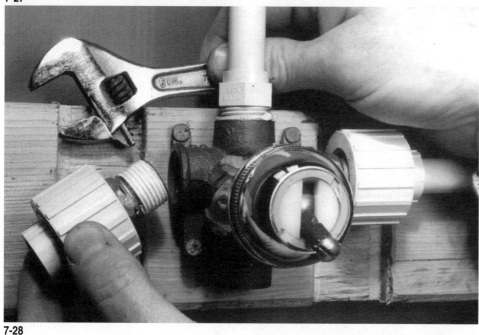

7-28

Plastic MIP threads. Male threaded solvent welding adapters may be used for nonpressurized or cold-water locations only. Genova Universal Male Adapters are for the cold water side only. For pressurized hot water with CPVC, a transition type fitting must be used (see previous chapter).

Plastic MIP fittings shouldn't be tightened as much as metal-to-metal fittings or they may crack. To install one, use TFE plumber's tape on the male threads. Install the fitting and hand tighten, then add one or two turns, at the most, for a watertight connection. When fully tightened, a correctly installed MIP adapter should still have threads showing. MIP adapters may be used for connecting a CPVC relief line to a T & P water heater relief valve. Also, you may use them for connecting shower risers to their mixing valves.

Water Hammer Mufflers. Genova Water Hammer Mufflers (Part No. 530901) may be used to prevent damaging water hammer at all fixtures and appliances. Use mufflers on the hot and cold water sides for every fixture except a toilet. And be sure that dishwasher and automatic washing machine hookups get them.

For installation details on Water Hammer Mufflers, see Chapter 14.

Shower supply. Tub/shower mixing valves (or separate ones, depending on which you get) are adapted to solvent welding tubing with MIP transition unions (Part No. 530451). The ½″ male threads screw right into the ½″ female threads of most mixing valves (Fig. 7-28). Connecting the transition union is shown in Fig. 7-29. If you are using Genova Universal Fittings, thread a ½″ brass nipple into the mixing valve's hot tapping using TFE tape or pipe dope and adapt to it with a Part No. 543051

7-29a, b A thick elastomeric gasket inside each transition union fits between the metal and plastic sides to permit differential thermal movement without leaks (7-29a). The mixing valve is attached to its 2" x 4" support with screws through the mounting tabs. Water hammer mufflers may be used above or below the mixing valve in any position, upright or not. They are pushed on and hand tightened.

7-29a

7-29b

Female Adapter. (Male Universal Adapters should not be used with hot water.)

In solvent welding CPVC, you can pipe upward from the shower mixing valve to the shower arm outlet starting with a simple MIP adapter (Part No. 50405). A transition union is not needed, because this isn't a pressurized connection. At shower arm height (see Fig. 5-9), install a Part No. 530551 Special FIP Wing Elbow.

It is fastened to a header (Fig. 7-31). A ½" CPVC tube connects the male adapter with the wing elbow, solvent welded, of course.

Later, during finish plumbing, the shower arm will screw into the special wing elbow, sealing against its elastomeric washer. No thread sealant is used. Fig. 7-32 shows a typical installation, with CPVC supply tubing.

Because a bathtub spout is subject to use as a hand grab and may receive

7-30 Here's a shower installation going in with ½" CPVC supply tubes that simply stab in and hand-tighten to the Genogrip Adapters.

7-31 The handy wing elbow fastens to the shower arm's support header with two screws. It accepts a solvent welded ½" CPVC shower riser coming from below and a ½" threaded chromed shower arm installed later during finish plumbing. It's another example of the way Genova Fittings help you get it done.

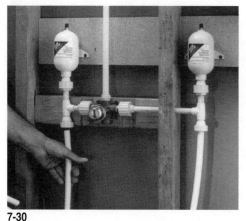

7-30

7-31

Tub/shower water supply hookup

½" Water Hammer Muffler 530901

Special ½" FIP Wing Elbow 530551

½" Water Hammer Muffler 530901

½" MIP Transition Union 530451

½" Male Adapter 50405

Mixing Valve

½" Tubing Hanger 52105

½" tubing hanger 52105

½" Tee 51405

½" Tee 51405

½" MIP Transition Union 530451

2' x 4' header for support

½" CPVC tube 50005

½" brass nipple

½" Tubing Strap 52115

½" CPVC tube

½" tubing hanger 52105

½" brass nipple for tub spout

½" tubing hanger 52105

7-32

quite a bit of abuse, I recommend doing the tub faucet hookup with threaded-brass nipples and a 90° brass elbow. This is shown in Fig. 7-32. You can purchase the brass parts from your dealer, selecting the lengths needed. Brass resists corrosion and eliminates the possibility of rusty water, though it costs more than other metal piping. Since the runs are short, cost is a minor factor.

Water testing. The ½" stubout pipes that extend through the wall to serve fixtures should be temporarily closed with solvent welding caps (Part No. 50155). Or else the fixture shutoff valves should be installed.

When the water supply system is finished and every opening turned off or else capped, it can be water tested under pressure. Go around looking for leaks. (Call for an inspection, if required.) After water testing and once the wall material has been placed over them, the solvent welded caps will be cut off and fixture supply

valves installed in their place.

Sillcocks. You can make sillcocks, through-the-wall hose taps, in either of two ways. In a cold climate use freezeproof sillcocks. In a mild climate use a lower-cost Part No. 530941 Genova Sillcock (see Fig. 14-6).

System drain. During your installation also keep in mind the possible need for draining the piping. It's a wonderful idea to have everything sloped slightly back toward the main house shutoff valve. Then if it's a Part No. 530271 3/4" Gate Valve with removable couplings, the system can be drained there.

Properly installed, your CPVC water supply system is a lifetime investment.

Shower is special

Because a shower or tub/shower is where a flow problem would be most noticeable, be sure you don't violate the no branches rule with it. Treat a shower with great deference.

"All plumbing fixtures should be on hand

before you begin."

Section 3 Doing Finish Plumbing
Chapter 8 How to Install a Bathtub/Shower

Bathtub Drain Connection

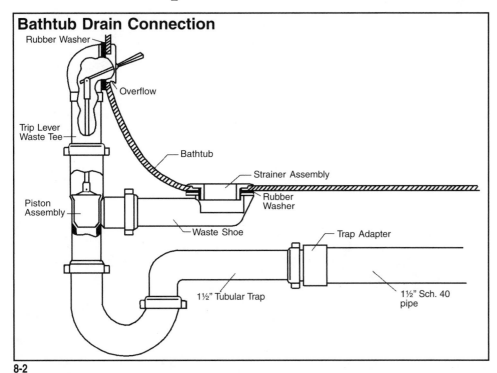

8-2

8-1

8-1 Finish plumbing consists of installing and connecting fixtures. Before this fiberglass tub/shower unit can be hooked up, the holes for its mixing valve and shower head must be bored. An electric drill makes quick work of it. Masking tape on the surface prevents chipping.

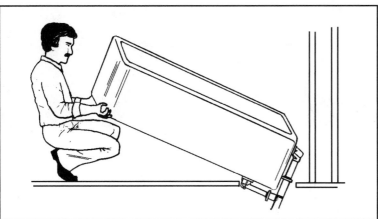

8-3

With the water supply system completed and the house walls closed in, you're ready to tackle finish plumbing. All plumbing fixtures should be on hand before you begin.

Bathtub installation

The first fixture to go in is the bathtub. Start by reading any instructions that came with it. These will pertain to providing proper as well as caring for the tub support during installation. One thing, the tub wall opening should be an exact fit for the tub. If it's too large, you'll have to frame it in at the foot end of the tub to close the gap. Moreover, you'll want to buy your

tub with its drain opening placed correctly for your installation, left- or right-hand.

If the tub is heavy, you may need help getting it into place, especially if it is to go on a second story. One way to handle a heavy bathtub without lifting it is to "walk" it across the floor, while still crated.

New metal tubs come with two holes: one for the overflow, one for the drain. The holes for fiberglass tub/shower enclosures need to be drilled out to fit your faucet's configuration (see Fig. 8-1). You'll need the trip lever drain hardware. Fig. 8-2 shows the parts arrangement.

Install the Genova trip lever overflow and drain fittings (Part No. 19015) in the tub openings with a ring of plumber's putty between them and the fixture. Once the jamnuts have been tightened down, the excess putty can be removed. Don't tighten too much or you could damage the tub's finish--"snug" is the word for it. The installation should be made so the drain doesn't support the weight of the tub.

Stand the tub upright at the head end of its enclosure. Gently lower its foot end to the floor (Fig. 8-3). The tub drain, should lower neatly into the hole you've created for it in the floor.

8-4

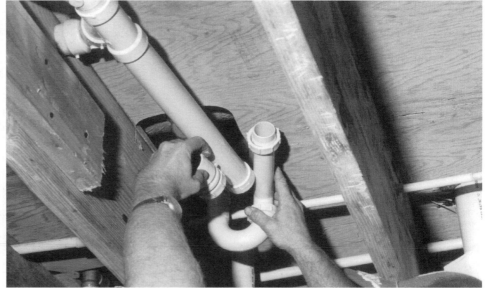

8-5

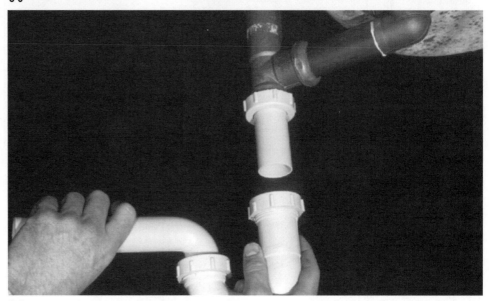

8-6

8-4 This Genova bathtub trip lever waste setup (Genova Part No. 19015) is made of polypropylene to surpass metal products. It fits to the tub's drain and overflow openings and has a tailpiece that fits the tub's trap.

8-5 Before sawing off the capped bathtub waste pipe, hold the P-trap and trap adapter in place to see how much of the pipe needs to be removed. A first-floor tub is accessed from the basement or crawlspace. A second-floor tub needs an access panel.

8-6 With the bathtub and its drain hardware in place, a 1½" tubular P-trap fits the tailpiece coming down from the trip lever waste.

Cast iron tubs either need no side support or they rest on flange support boards. Stamped-metal and fiberglass tubs should be fastened to the wall; the furnished instructions will tell how.

Connect tub drain. Now, working from the basement or crawlspace, or through an access door made at the head of a second-story tub (or else beneath it on the first floor), make the drain connection. Start this by marking and sawing off the tub's capped waste pipe to the proper length (Fig. 8-5). Then, install the tub's solvent welded trap adapter, which accepts its P-trap.

A 1½" tubular P-trap is the modern way of connecting a tub drain. This is Genova Part No. 175151 and uses a slip jamnut and washer to join the trap adapter watertight (See Fig. 8-4). For more about tubular goods, see

Chapter 11. If you prefer an old-style drum trap, you can get a PVC one (Part. No. 75715) to match the DWV pipe size.

Finishing the job. The wall next to the tub can now be tiled, if that is to be done. Joints between the wall and bathtub should be sealed with a good bathtub caulk with the tub about half full of water. Drained, it will rise slightly, compressing the caulking for a tight-fitting job. Then, build the access cover panel and install it. Finish off by installing the tub drain and its actuator along with the spout, faucet escutcheon(s), and handle(s). Also install the shower arm, escutcheon, and shower head.

Run some water into the tub, drain it, and check for leaks at the trap and drain connections. Further tightening beyond hand-tight should cure them.

8-7

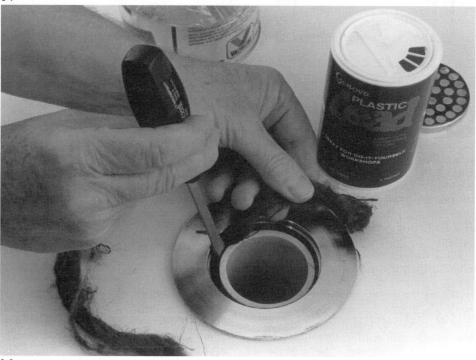

8-8

8-9

8-10

8-7 Shower drains get caulked joints that allow them to be removed without access from beneath. Rope oakum is packed densely into the joint between the shower waste pipe and shower drainage fitting using a hammer and plumber's corking tool. Pack to within 1″ of the top.

8-8 If you don't have a corking tool, a screwdriver can be used for installing the oakum packing. Once the oakum is packed tightly, about an inch of the waste pipe should be left exposed.

8-9 Dry Genova Plastic Lead powder is first mixed with water to a putty consistency. Then fill to the top of the pipe with the caulking lead substitute. A screwdriver makes a decent "trowel." Plastic Lead expands as it sets to protect the oakum.

8-10 The last step is to install the drain grating. The use of Plastic Lead saves having to melt and pour plumber's lead to make a caulked joint. However, you can do that if you prefer.

Shower installation

A shower by itself comprises a molded shower base and upright panels that fit it. You can buy a shower stall complete with upright panels and molded base or build your own shower using waterproof backer-board, tiling over it. At any rate, you still need the molded base (way to go, for us DIYers). Choose the size for the space you have. A metal shower drain fitting should be installed in the base drain opening, putting a ring of plumber's putty around inside the base. Run the locknut on from the other side.

Position the base directly over the shower's waste pipe, drain fitting centering around the pipe. Mark the waste pipe for saw-off a little below the top of the drain fitting. Lift the base off and make the cut. Thus when you set the base down over the pipe again, the pipe will be slightly short of reaching flush with the top of the fitting. While you're at it, make deep score marks with coarse sandpaper all around the upper 2″ of the waste

**Caulked Shower Drain
(cross section)**

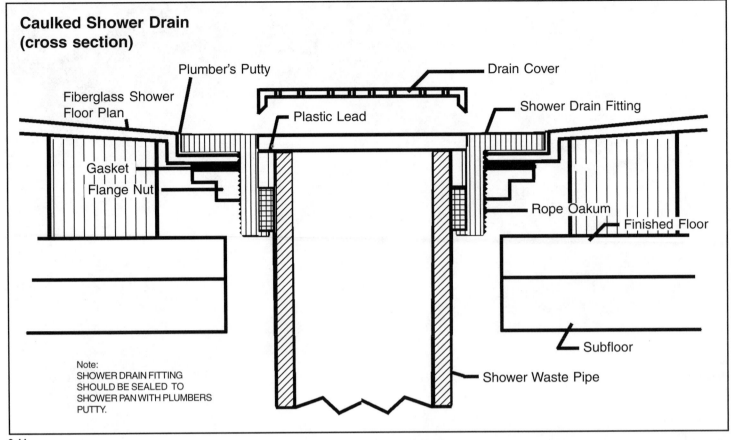

Note:
SHOWER DRAIN FITTING
SHOULD BE SEALED TO
SHOWER PAN WITH PLUMBERS
PUTTY.

8-11

pipe. Later, the roughness will help tie the shower base and waste pipe together, eliminating movement between them.

See that the base is well supported, especially in the center. It **must not** depend on the waste pipe for support or else the joint is bound to move and leak. Stand in the center of the base. It should not deflect under your full weight.

Caulking the joint. A shower-to-waste pipe joint gets caulked. This kind of joint is made by filling the gap between the pipe and drain fitting with a unrotting water-impervious material. A caulked joint lets the shower be removed without having to destroy the plumbing system to get it out. To make caulking easy, the shower drain fitting contains a deep, flange-bottomed recess that's about ¼" larger than the outside diameter of a 2" waste pipe. The space between the pipe and fitting is caulked.

Caulking consists of packing plumber's rope oakum for half the depth, then covering it with a layer of Genova Plastic Lead (Part No. 14210) for the second half. Oakum-packing goes best if you have a plumber's corking tool (see Fig. 8-7). Since you probably won't want to purchase the tool for just one shower, you can work with a screwdriver (Fig. 8-8) and

hammer. Pack hard enough that the oakum alone will seal the joint. Old as Roman times, oakum swells on contact with water to help it seal.

The purpose of Plastic Lead is to hold the oakum in place. Do it as shown in Fig. 8-9. Fig. 8-11 shows a cross section of a caulked shower drain.

Shower walls. If you build your shower's walls, use waterproof backer-board reaching down inside the shower base to ¼" above the molded-in curb. That joint will later be filled with bathtub caulk to keep water going down the shower drain, not onto the floor behind the walls. The wall material should reach up at least 60". Your home-built shower walls can be tiled or covered with another material approved for such use.

Chapter 9 Other Uses for Plastic Lead

The good folks at the Genova spread developed Plastic Lead as a scientifically formulated caulking lead substitute. Little did they know that before long a whole lot of other uses would be found for it. Nonplumbing uses. Some are illustrated on these pages.

Plastic Lead (Genova Part No. 14210) comes as a ready-to-use off-white powder packed in air-tight shaker cans.

It can be sanded, carved, drilled, even tapped. Plastic Lead develops a strong, water resistant bond to most porous surfaces, including cast iron, clay tile, concrete block and brick—even gypsum board and wood. Its free-flowing properties enable it to fill fine crevices and its surface is readily troweled or wiped smooth. It can be painted without special preparation. Amazing stuff!

Genova Plastic Lead is water resistant because it contains precatalyzed melamine formaldehyde resin, which is activated by water. This is the principal ingredient of unbreakable, dishwasher-cleanable melamine dinnerware.

The cure-without-shrinking ability of Plastic Lead is most unusual. Practically everything that's mixed with water and sets hard, including plaster and concrete, shrinks on setting. Some shrink considerably. But, as long as it is not mixed too wet, Plastic Lead will not shrink on curing. This makes it ideal for such things as grouting tile joints, filling cracks, and plugging holes.

If you ever need a material around the house that fits these specs, try good old Plastic Lead.

Genova Plastic Lead has been used for such diverse applications as bedding and filling cracks in marble and as a crack filler for wood, stone, and concrete. It is highly suited to setting and bedding toilets to concrete floors. These are only a few of the many applications for versatile, water-resistant Genova Plastic Lead.

9-1 Genova Plastic Lead comes in a handy-to-use shaker can. Open the lid by twisting the cover and shake out the amount you need. Mix with water as directed on the can. More or less water may be used to form the desired consistency—thin like this for faithfully reproduced castings or thicker for patching and as a caulking lead substitute. Reclosing the cover preserves the rest of the Plastic Lead for subsequent use.

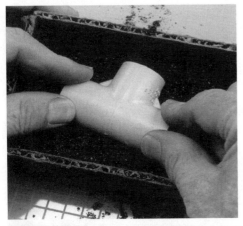

9-2

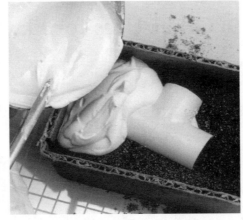

9-3

9-4

9-2 To make a sand-embedded casting with Genova Plastic Lead, make a frame with corrugated cardboard. Fill it partially with sand, then push the object to be cast—here a CPVC tee -halfway into the sand.

9-3 Pour Genova Plastic Lead caulking lead substitute over the object, leveling it even with the top of the form. It flows into place, and picks up fine detail. Before the Plastic Lead sets, the casting's back received an embedded wire-loop hanger.

9-4 Shortly the mold was stripped, leaving the Plastic Lead wall plaque neat as you please. The natural cream color of Plastic Lead shows through between grains of embedded sand.

9-5 Like any faithful casting material, Plastic Lead takes on the surface texture of whatever it's cast against. To make this desktop "flying saucer," Plastic Lead was cast into a plastic funnel. After setting, three holes were drilled into the bottom and self-threading Raingo® Screws driven in as "feet." The shiny plastic surface of the funnel was faithfully transferred to the casting.

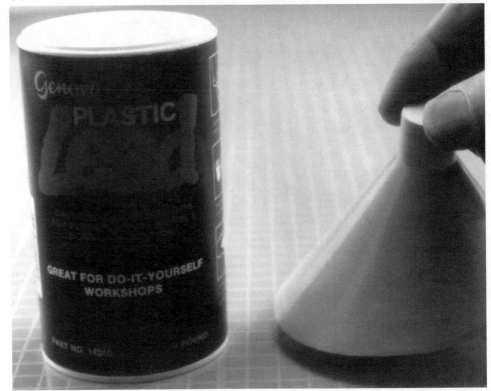

9-5

Mixing. Because Genova Plastic Lead wastes no time in setting, mix only what you'll be using in 10 minutes or so. Shake out some powder, add a little water, and mix. Something like 2½ ounces of water should be added to a half pound of Plastic Lead. When adding water, it's best to put in half the water right away and mix. Then add the rest of the water a little at a time, mixing between additions, until the desired consistency is reached. This will keep you from getting the mix soupier than you want it. It's best to keep the mix as thick as possible, provided that you can work with it easily. For hole and crack filling, you want it fairly thick. For casting, where you want it to reproduce delicate details, use more water to make a soupier mix.

Open working time varies from 10 to 30 minutes, depending on how thick it is when used. In 24 hours Plastic Lead hardens fully into a stone-like product of extreme hardness and durability.

Applying. A screwdriver blade or putty knife makes a good applicator for Plastic Lead at putty consistency. For pouring, you can use a cup or can. The harder you can pack Plastic Lead in, the denser it will become. This is another reason for using relatively stiff mixes. Still, you don't want the mix so dry that the material crumbles.

One thing to remember: **non-plumbing uses for Plastic Lead are experimental.** Don't use it for anything where a failure could injure someone. For example, anchoring a

chandelier with Plastic Lead or patching a broken electrical appliance with it wouldn't seem wise. And don't depend on Plastic Lead as a glue or adhesive. It is not. Stick to appearance uses, and especially to using Plastic Lead as a handy caulking lead substitute.

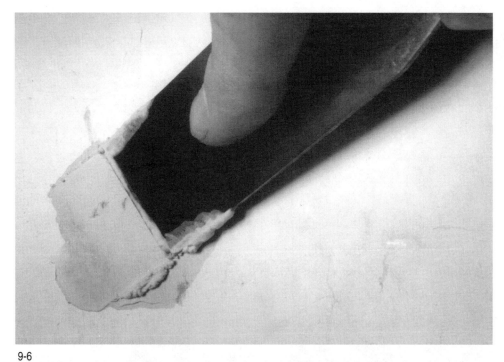

9-6

9-8

9-7

9-9

9-6 A great use for Genova Plastic Lead is filling holes in walls, large and small. With it, one filling flush with the surface does the job. There's no need to go back several times to fill sunken spots where the filler shrinks. Apply Plastic Lead with a putty knife. Any excess can be sanded off after setting.

9-7 Good old Plastic Lead is terrific for patching holes in masonry and stucco walls left when masonry anchors have been removed. Trowel it in with a putty knife, then rub over it to texture like the wall.

9-8 Another terrific use for Plastic Lead is filling around loosened wood screws. Just dip the screw in Plastic Lead before driving it. Wipe off any excess that oozes out. Once the Plastic Lead sets, the screw should stay put until removed.

9-9 You'll really like Plastic Lead as a tile grout, too. It works as easily as anything you've used, but doesn't shrink and leave cracks. Mix to spreading consistency and apply with a squeegee or sponge, working it well into joints between tiles. A little later, clean up with a damp sponge. As with other grouts, the haze that remains can be wiped off easily with a dry cloth.

9-10 Plastic Lead sands wonderfully, every bit as easily as wood putty. What's more, it takes paint well. When staining Plastic Lead, as with wood putty, you shouldn't count on it to match the wood. Experiment first.

9-11 Shrinkage comparison between Genova Plastic Lead (top) and wood putty (bottom) shows the Plastic Lead plug holding snugly, while the wood putty plug has shrunk and come loose.

9-12 Can you drill Plastic Lead? You bet! Plastic Lead loves to be drilled. It doesn't clog the bit, or have any grain to deflect it off target.

9-13 Incredibly, you can tap a pilot-drilled hole in Plastic Lead and install a machine screw. Such a connection could be used for light anchoring chores, but shouldn't be counted on for heavy duty holding.

9-14 By dropping a machine screw or bolt into a bored hole in concrete and filling in around it with Plastic Lead, you can manage light anchoring in concrete floors and walls. Resistance to pull out and shear will amaze you for such an easily used product. Don't, however, stake your safety on it.

9-10

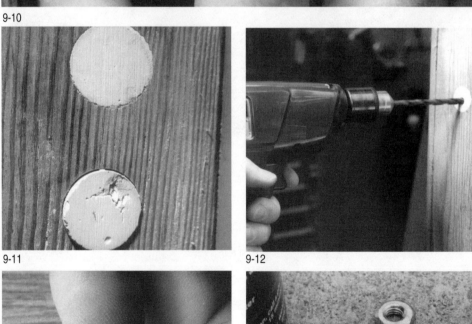

9-11

9-12

9-13

9-14

Chapter 10 Poly Risers Make it Easy

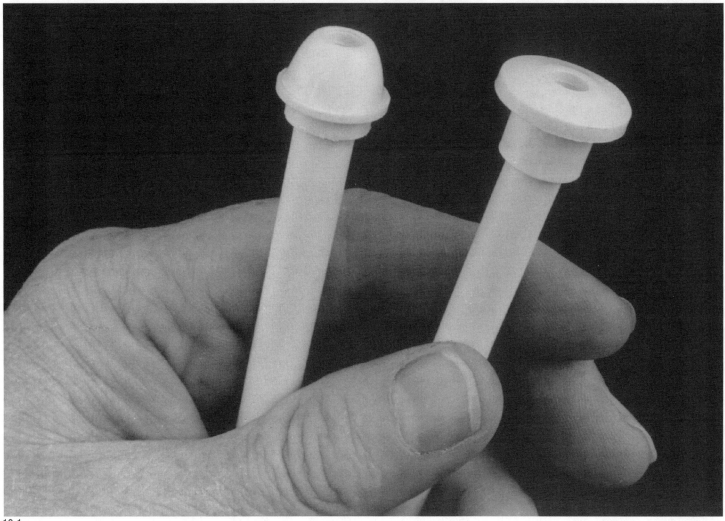

10-1

Washbasins, toilets, sinks, laundry tubs, and dishwashers are easiest to join to their hot and cold water supply piping with Genova's noncorroding, flexible supply tubes called Poly Risers. Made of 3/8" O.D. flexible thermoplastic tubing, Poly Risers are so flexible you can even tie one in a knot without breaking or kinking it.

Poly Risers come with two choices of preformed ends (Fig. 10-1). Both types make for a quick, easy, leak-free connection.

Besides being excellent for new plumbing, Poly Risers are terrific for replacement use.

What's available. A Poly Riser hookup begins with a pair of fixture supply valves at the wall or floor behind the fixture. Those by Genova are easiest to use, because they contain push-in-hand-tighten Genogrip fittings at both ends. The 1/2" Genogrip at the inlet end lets the valve be installed

over any 1/2" tube (5/8" O.D.), whether CPVC or copper tubing. No sweat soldering or solvent welding is needed. The valve's outlet has a 3/8" Genogrip that accepts a Poly Riser (or other 3/8" O.D. riser tube in plastic or metal).

Both angle and straight supply valves are offered by Genova (Fig. 1-7). The angle valve is for use with a water supply coming from the wall (Genova Part No. 530651). The straight supply valve is for use with a floor water supply (Part No. 530301). Either valve provides vital water control as well as easy fixture connection.

Poly Riser tubes are made in 12" and 20" lengths plus an added 36" length for use with washbasins and sinks having floor water supplies. Poly Risers work both with Genova supply valves and those of other brands. In this use, they make a direct

10-1 Two types of Poly Risers let you simplify the hookup of a fixture to its water supply. The bullet-nosed Poly Riser (left) is for connecting washbasins and sinks. The flat-ended Poly Riser at the right is for connecting toilets. A 1/2" faucet nut does the coupling.

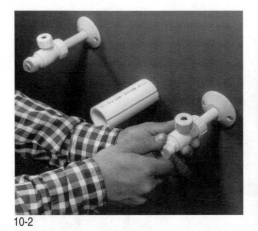

10-2

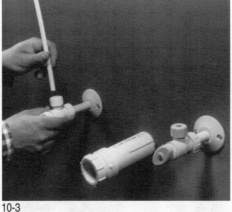

10-3

10-4

10-2 To prepare for a fixture's water supply connection, install Genova supply valves at both hot and cold water stubouts. These slip on and hand tighten over the 5/8" O.D. stubs, whether they're CPVC or copper tubing.

10-3 Then push the lower end of a Poly Riser into the 3/8" Genogrip fitting on the supply valve. When it bottoms out inside, hand tighten the 3/8" collar.

10-4 Then couple the upper shaped end of the Poly Riser to the faucet's tailpiece using a 1/2" faucet nut.

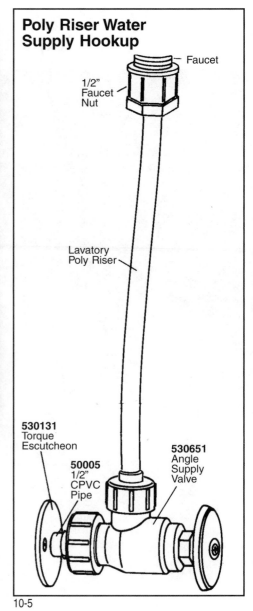

Poly Riser Water Supply Hookup

Faucet

1/2" Faucet Nut

Lavatory Poly Riser

530131 Torque Escutcheon

50005 1/2" CPVC Pipe

530651 Angle Supply Valve

10-5

replacement for old-style 3/8" O.D. metal risers. Fig. 10-5 shows a Poly Riser water supply hookup with a fixture supply valve.

Using Poly Risers. The plain end of a Poly Riser goes into the supply valve's 3/8" O.D. Genogrip fitting. The shaped end installs upward toward the fixture.

Much of a Poly Riser installation is the same for all uses. First cut the tube to length. Then apply liquid soap, or Genova All Purpose Liquid Seal Lubricant to the Poly Riser's lower end. Bend it in a smooth curve and push it into the supply valve's Genogrip adapter. If it helps, you can loosen the supply valve's 5/8" O.D. Genogrip fitting and rotate the valve to face for the smoothest curve. Be sure to retighten it. See that the Poly Riser is pushed in all the way, bottoming on the inner shoulder. Figs. 10-2 to 10-4 show how to use Poly Risers. Instructions also come with these carded products.

Hookups

Poly Riser's variety in sizes suits whatever use you might have for them.

Lavatory/sink. In most cases 20" Poly Risers (Part No. 530861) will reach from a wall supply valve. Use 36" (Part No. 530881) with a floor supply,

12" (Part No. 530841) if the supply valve and faucet are really close.

How you connect Poly Risers to a faucet depends on the configuration of your faucet's tailpieces. On older sink faucets having removable threaded brass tailpieces, discard the tailpieces and connect the Poly Risers directly to the faucet with the same 1/2" faucet nuts that held the brass tailpieces.

The "no tools" and "do-it-yourself" faucet hookups are made in the same way but with nothing to be discarded. Fig. 10-7 shows making one such tailpiece hookup. Connections up behind the bowl are best made with a basin wrench (Fig. 10-6).

To use Poly Risers with faucets having plain-ended 3/8" O.D. copper tubes, you have a choice of methods. You can cut off the shaped Poly Riser end and flare both the riser and the

tailpiece. (The flare nuts should be installed first.) Then join the two with a 3/8" O.D. flare coupling. Genova Universal couplings may be used instead of flare couplings. That way, you won't need a flaring tool. Fig. 10-9 shows different Poly Riser-to-faucet hookups.

Toilet hookups. In most cases a 12" long Poly Riser (Part No. 530801) will reach a toilet tank. If it won't, then a 20" (Part No. 530821) surely will.

Lengthening a Poly Riser

If a Poly Riser isn't long enough to reach, you can easily lengthen it with a Genova Part No. 541031 Universal coupling and another Poly Riser. It has Genogrip adapters at both ends to accept the plain ends of Poly Risers. Before leaving the job, check for leaks with the water turned on.

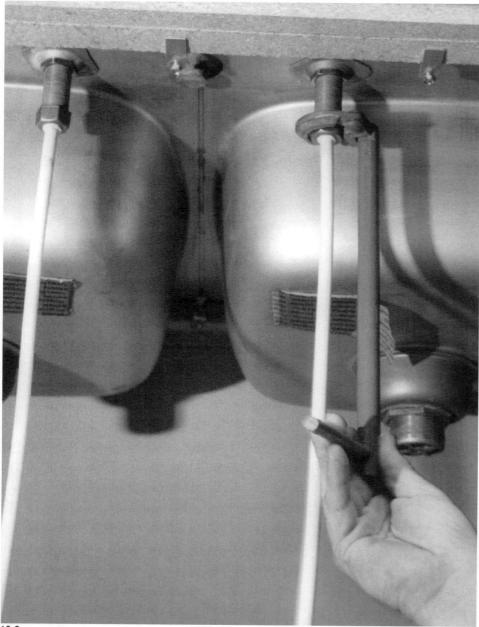

10-6

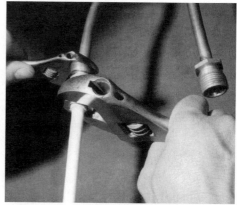

10-7

10-8

10-6 Connections located up behind fixture bowls, as shown in this view from behind the fixture, can be tightened comfortably using a basin wrench. The wrench jaws are positioned to tighten or loosen, depending on which way you face the movable jaw.

10-7 When connecting to faucet tailpieces with "do-it-yourself" ends like this, be sure to back up the other side of the joint with another wrench to keep it from twisting, which could cause a leak.

10-8 Poly Risers also may be connected with compression fittings. Slip the ferrule over the end of the Poly Riser to make sure it's a snug fit. If not, apply a little silicone rubber sealant (RTV) between the ferrule and Poly Riser to seal against leaks.

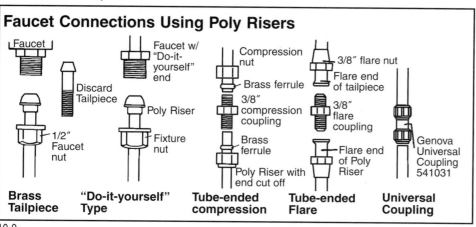

Faucet Connections Using Poly Risers

Faucet

Discard Tailpiece

1/2" Faucet nut

Brass Tailpiece

Faucet w/ "Do-it-yourself" end

Poly Riser

Fixture nut

"Do-it-yourself" Type

Compression nut

Brass ferrule

3/8" compression coupling

Brass ferrule

Poly Riser with end cut off

Tube-ended compression

3/8" flare nut

Flare end of tailpiece

3/8" flare coupling

Flare end of Poly Riser

Tube-ended Flare

Genova Universal Coupling 541031

Universal Coupling

10-9

Do-It-Yourself Plumbing...

It's easy with Genova

Chapter 11 Finishing Bath Plumbing

11-1

11-2

11-1 **Before.** A bathroom before remodeling featured the old fixtures and fixture arrangement there when the house was built.

11-2 **After.** When the bath was remodeled with Genova CPVC water supply and PVC-DWV systems using new fixtures, the locations of toilet and washbasin were exchanged. Its usefulness and appearance were greatly improved.

Finish-plumbing a house with Genova drain-waste-vent and water supply systems is simple and conventional. Whether used for new construction, for an add-on bathroom, or for a bath remodeling, Genova materials may be used right up to the fixture drains and faucets.

Start the finish plumbing by mounting the washbasin (Fig. 11-3). Instructions for doing this usually come with the basin or vanity. How to handle a bathtub and shower is

covered in Chapter 8. Following is how to do the rest of the bathroom:

Toilet
Connecting a toilet consists of fastening the bowl with its self-contained trap down to the floor waste opening and installing its wall or floor water supply. Follow the instructions given with the toilet, when mounting the tank (if separate).

Toilet soil pipe. A toilet's soil pipe run starts at the floor with a Genova

"Pop-Top" toilet flange (this is described in Chapter 4), which comes in several configurations. Which one to use depends on what type of floor you have: wood or concrete slab. Either way, the toilet flange centers 12" out from the finished bathroom wall. If not already done, solvent weld the toilet flange to the soil pipe beneath the floor. (See Figs. 11-4 to 11-6.)

Then fasten the flange to the floor (Fig. 11-7) and knock the "Pop-Top"

11-3 In finish-plumbing a bathroom, the first step is to install the washbasin. This countertop unit merely sets down on its vanity base. The single-handle faucet has been mounted to the top, as this goes easiest while the countertop is off.

11-4 For connecting a toilet, solvent weld a Genova "Pop-Top" toilet flange to the below-floor soil pipe. But first, clean the mating parts with Novaclean®. Then coat the outside of the pipe generously with Genova All Purpose solvent cement (or Novaweld® P). The floor opening should be 5" diameter to clear the shoulder of the flange.

11-5 Next, spread solvent cement on the inside of the PVC toilet flange. These steps may well have been done earlier as part of the rough plumbing to close off the DWV system for water-testing.

11-6 Immediately push the toilet flange down over the soil pipe with a slight twist. If the outer ring is not a free-wheeling one like this one is, quickly bring the bolt slots into alignment so that a line drawn through them would be parallel to the wall behind the toilet.

11-7 Then you can screw the flange to the floor with four No. 10 1½" flathead wood screws. The bottom of the toilet flange should rest on the finished floor. The toilet will later bolt to the flange for a secure mounting using special toilet mounting bolts. The flashed-over drain opening seals the DWV system until just before the toilet is set.

11-8 A few light taps with a hammer will knock out the "Pop-Top" flashing. Do this just before installing the toilet bowl and while wearing safety glasses. Remove the disc and discard it.

11-9 Patented serrated slots designed into the Genova toilet flange grip the hold-down bolts, keeping them upright. Slide the bolts around until they are equally distant from the wall behind the toilet.

11-3

11-4

11-5

11-6

11-7

11-8

11-9

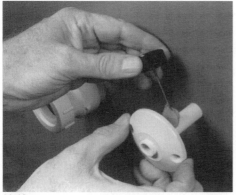

11-10

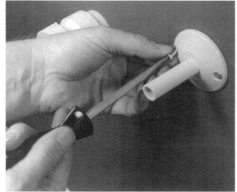

11-11

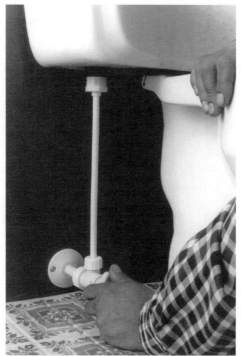

11-12

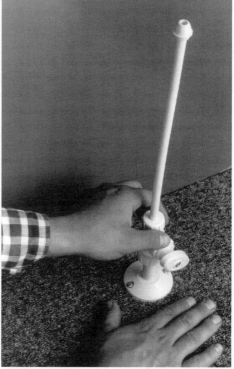

11-13

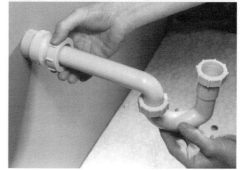

11-14

out with a hammer (Fig. 11-8). Then place a toilet hold-down bolt (a standard hardware item) in each of the serrated slots of the flange (Fig. 11-9).

Unpack the toilet bowl and rest it upside down on a thick padding of newspapers to prevent damage to it. The finished floor material should already have been installed and the bathroom wall finished. Installation of a toilet is described in Chapter 13.

Toilet water supply. Now you're ready to connect the toilet's water supply to its tank inlet. The capped ½" CPVC toilet water supply stubout left from rough-plumbing and pressure-testing for leaks can be cut off, leaving about 2" sticking out from the finished wall or floor. Solvent weld a Special Torque Escutcheon (Genova Part No. 530131) over the stubout (Fig. 11-10) and attach it to the wall or floor with a pair of screws (Fig. 11-11). As soon as the solvent cement has set a little, you can install a Part No. 530651 Genogrip Angle Supply Valve (with a wall stubout—see Fig. 11-12) or a Part No. 530301 Genogrip Straight Supply Valve (with a floor stubout—see Fig. 11-13). Both of these handy valves incorporate Genogrip Adapters that "look" upward toward the toilet. Both accept Poly Risers and fit ⅝" O.D. CPVC or copper stubout tubes in a Genogrip connection. The ½" tubes measure ⅝" O.D. No lead-free sweat-soldering and

no solvent welding are required.

Once installed, the toilet is ready to test-flush. It's satisfying to watch the water flow through the bowl and disappear down the drain just the way it is supposed to and know that you built the system.

Washbasin

The washbasin bowl, drain, pop up, and faucet should be installed following the specific instructions that come with them. Use plumber's putty under both sides of the bowl's rim, around the drain inside the bowl, and beneath faucets that don't have their own base gaskets. The outside of the washbasin drain assembly usually gets a thick elastomeric gasket. This readies you for making the washbasin waste connection.

Washbasin waste. You'll note that the washbasin's 1¼" diameter drain tailpipe has very fine running threads at one end. (Running threads are not tapered like pipe threads.) No one ever tells you, but you should put some pipe joint compound on these threads to keep them from leaking. Putting a little silicone rubber sealant (RTV) around them before screwing the tailpipe into its drain fitting will have the same effect. That done, hand-tightening of the tailpiece is plenty.

Trap adapters. The washbasin—also a sink or laundry tub—connects to its 1½" PVC wall or floor waste pipe

11-10 To relieve the twisting force exerted on a CPVC tube stubbed through a finished wall or floor, use Genova's Special Torque Escutcheon. First solvent weld it to the stubbed out tube.

11-11 Then anchor it to the wall or floor with two screws.

11-12 When the toilet is installed—both its waste and water supply—you can turn on the supply valve. Having the water supply stubout as low as possible on the wall lets the angle supply valve be low, too. This way, the toilet's Poly Riser can be long as possible, making for an easy connection.

11-13 If supply valves must be located in the floor beneath a fixture, use a Genova Straight Supply Valve that has a Genova Adapter. Used at floor or wall, these valves and Poly Risers make an unbeatable combination for water supply hookup.

11-14 Genova is the only firm making 1¼" traps in noncorroding, heat resistant, chemical resistant polypropylene.

11-15 All Genova trap adapters come with dual-sized washers to fit either 1¼ " or 1½ " tubular products. They're called the "Dynamic Duo." These two figures show how to use "Dynamic Duo" washers. "Dynamic Duo" washers will accept either metal or plastic traps. The "Dynamic Duo" assembly slips on over a 1¼ " washbasin trap's arm, with the slip jamnut going on first. Then the trap can be fitted snugly to a 1½ " trap adapter.

11-16 Used with the lighter-colored ring astride the darker one, the washer fits smaller 1¼ " washbasin traps (top left). For use with 1½ " traps, the lighter-colored outer washer is used alone, as shown bottom left.

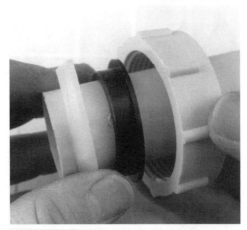

11-15

11-16

with Genova polypropylene **tubular goods**. (For more about tubular goods, see the next chapter.)

To make tubular goods fit the DWV system, a trap adapter is needed. A Part No. 72211 trap adapter solvent welds over the 1½" PVC washbasin waste pipe coming from the wall (or floor). For this reason, it's called a **pipe** trap adapter. Of course, you'll first want to saw off any water test cap installed there leaving about 4½" of waste pipe protruding from the wall. A Part No. 72311 trap adapter solvent welds inside a 1½" fitting socket behind the wall. This is called a **fitting** trap adapter. To help you plumb, Genova markets both kinds.

Instructions in the next chapter tell more specifically how to install a trap. Washbasins with wall waste pipes get 1¼" tubular P-traps (Fig. 11-14). P-traps are called that because of their shape. The size increase—from 1¼" to 1½"—is done with a "Dynamic Duo" washer at the wall. It comes with the Trap Adapters. Washbasins with floor waste pipes get 1¼" S-traps. The size increase, then, is done at the upper end of the S-trap, again using a "Dynamic Duo" washer.

The 1¼" P-trap of choice is Part No. 175141. The Genova S-trap of choice is Part No. 177141. It's also made of polypropylene. It is always best to serve a washbasin with a 1¼" trap,

rather than a 1½" trap. This is because the larger 1½" traps tend to be "lazy" when installed on washbasins — they get too little flow to purge them of trap-restricting debris. Genova makes both 1½" and 1¼" P-traps and S-traps, but on washbasins, it's best to use the 1¼".

What you must **not** do is go from a larger size to a smaller size drain part, such as from a 1½" trap to a 1¼" waste pipe. This violates an old plumber's axiom for drainage piping.

Washbasin water supply. Genova supply valves—angle valves for wall supplies and straight valves for floor supplies—help you connect to the washbasin's faucet. Flexible Poly Risers reach from the supply valves up to the faucet tailpieces.

Once the fixture has been mounted, and its waste and water supply hookups made, the water can be turned on. Run out a bowl of water and drain it, looking for leaks in the tubular drain piping. If you find a leak, further tightening of the slip jam-nut usually cures it.

Then you can sit back and feel good about having completed a successful home plumbing project.

Chapter 12 Finishing Kitchen Plumbing

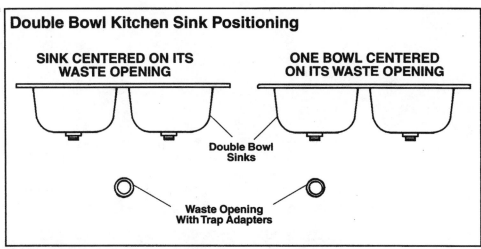

Double Bowl Kitchen Sink Positioning

SINK CENTERED ON ITS WASTE OPENING

ONE BOWL CENTERED ON ITS WASTE OPENING

Double Bowl Sinks

Waste Opening With Trap Adapters

12-1

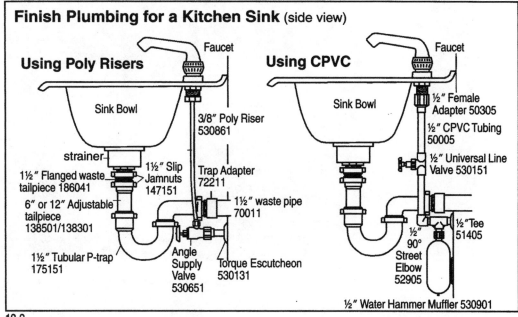

Finish Plumbing for a Kitchen Sink (side view)

Using Poly Risers

Faucet

Sink Bowl

strainer

1½" Flanged waste tailpiece 186041

1½" Slip Jamnuts 147151

Trap Adapter 72211

3/8" Poly Riser 530861

6" or 12" Adjustable tailpiece 138501/138301

1½" waste pipe 70011

1½" Tubular P-trap 175151

Angle Supply Valve 530651

Torque Escutcheon 530131

Using CPVC

Faucet

Sink Bowl

½" Female Adapter 50305

½" CPVC Tubing 50005

½" Universal Line Valve 530151

½" 90° Street Elbow 52905

½"Tee 51405

½" Water Hammer Muffler 530901

12-2

Doing the finish plumbing for your kitchen consists of installing the sink. The double bowl kitchen sink is fairly standard and should be positioned for best use within the kitchen's floor plan. During rough plumbing, its water supply and waste pipe should have been arranged to suit that location. Sink rough-in dimensions are shown in Fig. 5-8.

A double bowl sink is either centered itself on the waste pipe (Fig. 12-1, left) or one of its bowls is centered on the waste pipe (Fig. 12-1, right). If the sink gets a food waste disposer, this is usually placed in the right hand sink bowl. In that case, the wall-waste pipe should either center on the left-hand bowl or center

between the bowls. (How to install a food waste disposer is covered in Chapter 14.)

Fig. 12-2 shows a side view of the parts involved in a sink's finish plumbing. The first step in a sink installation is to solvent weld the trap adapter (Fig. 12-3).

Mount sink. Very often, the sink is part of a counter base. It is then fastened in a cutout made in the countertop. The easiest way of mounting the faucet is to do it before placing the sink in its cutout. Some faucets are set in plumber's putty, while others have their own base gaskets. Install the spray hose, if there is one, in its sink opening by following all of the instructions by the

faucet and sink manufacturers.

Kitchen sinks use separately installed basket strainers. These are sealed into sink bowls with plumber's putty. A thick rubber gasket and flat washer go on the underside of the sink. The assembly is squeezed tightly to the sink bowl opening by a large jam nut, which threads to the basket drain. You can generally get it reasonably tight by hand, but to make it really tight, turn it further by pushing with a screwdriver or turn it with a special strainer wrench. Clean up the excess plumber's putty inside the sink before leaving that part of the job.

With the sink in place, you can connect its water supply to its faucet. The water supply is done first, so that

12-3 To finish the kitchen sink's plumbing, first cut off its capped waste pipe stubout 4½″ long and solvent weld a trap adapter (Genova Part No. 72211) to it. Genogrip Angle Supply Valves have been fitted to the cutoff ½″ water supply stubs.

12-4 If, instead of a waste stubout, you have a 1½″ fitting socket waste opening, solvent welding a fitting trap adapter (Part No. 72311) will enable it to accept a tubular P-trap.

12-5 Tubular drains use slip joints between parts. These are comprised of a slip washer and a slip jamnut to hold the washer tightly between the tube and its fitting. The fitting is threaded for the nut. A slip washer's flat side should always be installed facing the nut.

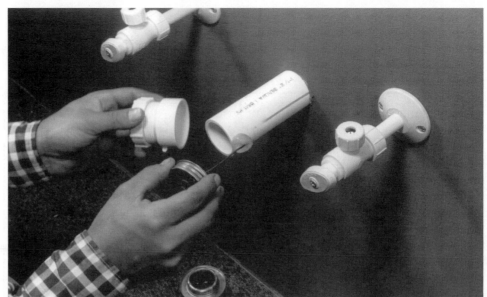

12-3

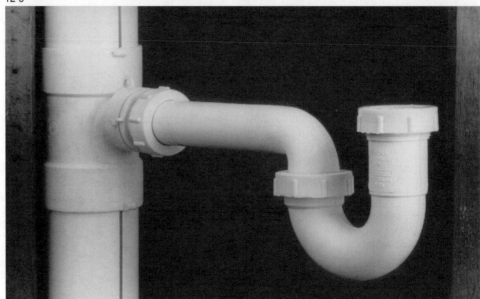

12-4

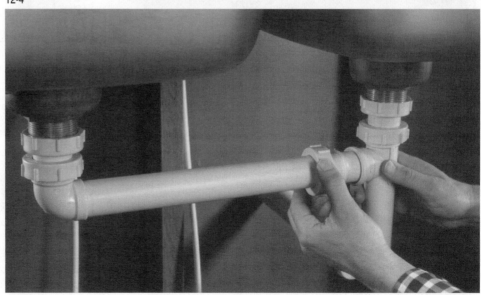

12-5

12-6

12-7a

12-7b

these parts will be out of the way when the drain parts are put in later. It goes just as described for washbasins in the previous chapter. With a wall supply, use Genova Part No. 530651 Genogrip Angle Supply Valves and Part No. 530861 20″ Poly Risers. These come two per pack—just the number you need for finish-plumbing a sink. If the sink water supply is from the floor rather than the wall, you'll need 36″ Poly Risers (Part No. 530881).

Tubular goods

The parts for hooking up your sink's drains to their waste pipe are called tubular products. This distinguishes them from the very different Drain-Waste-Vent piping. Tubular drainage products are made thinner walled than DWV. A tubular drain's walls are 0.062″ thick vs. 0.146″ for a DWV pipe—thus, DWV pipe walls are more than twice as thick as tubular walls. What's more, tubular parts use slip joints (Fig. 12-5) instead of solvent welded joints. Because tubular goods are thinner-walled than DWV, they must be made of a more heat resistant material. Genova's polypropylene is the best tubular material.

Tubular drainage products are not meant for use behind house walls. You must use them underneath fixtures where you have ready access for maintenance.

Why Polypropylene?

Having the thinnest walled pipes closest to the fixtures where they'll receive the very hottest water right from the fixtures is seemingly a backwards situation in plumbing. Even worse, they also get the most concentrated chemicals, such as drain cleaners. To make a tubular drain's life still harder, hot water and those caustic, heat producing chemicals don't flow quickly through your fixture's drains as they do in the house DWV system. Instead, they stand and soak in the low points of

traps...for minutes, perhaps hours.

Fortunately, Genova tubular goods are up to the torture, because they're made of a tough heat and chemical resistant plastic called polypropylene (PP). In fact, PP is so unaffected by chemicals that it cannot be solvent welded. That's why Genova chose polypropylene for its tubular products. Polypropylene, being a thermoplastic, does not rust, rot, or corrode. It is definitely the material of choice for tubular drain products.

Some tubular systems use PVC; it's a fine material, but not the best one for tubular products. Some PVC tubular systems have cheaper-to-make ABS slip jamnuts. Those are the threaded nuts that hold slip jointed tubular drainage parts together and keep them from leaking. Genova uses polypropylene slip jamnuts. Even if you accidentally drop some solvent cement on one, it won't be welded to its threads.

The point in bringing this up is that, in installing a Genova system, you can make use of sealants to prevent leaks in slip joint parts (Figs. 12-7a and 12-7b). But don't try this with non-Genova ABS slip jamnuts. They will fail from what plastics chemists call stress cracking. Stress cracking occurs when ABS is exposed to the oils in plumber's putty or the chemicals in silicone rubber sealant, two of the most common slip joint sealants.

So, if you leak proof a Genova or a metal slip jamnut, no problem. But with any other nut, don't try to leak proof with plumber's putty or sealants.

Meet tubular parts

Traps. The best known tubular drainage item is the trap. Most common is the P-trap. It has two parts, a J-bend and a longer waste arm. These are joined in the center by a swiveling connection (Fig. 12-8). The center joint parts are carefully fitted to let a trap be swung to the right or left to suit different fixture-

12-6 Pouring boiling hot water down a sink drain won't harm tubular drains if they're polypropylene. That's why this heat and chemical resistant thermoplastic is the material of choice for all tubular drains.

12-7, a & b Here's a trick that pro plumbers often use to avoid call-backs for leaks: run a little silicone rubber sealant (RTV) (12-7a) or plumber's putty (12-7b) around the inside of the face of the slip jamnut before making up the joint. These materials will not harm Genova Polypropylene Tubular Parts (see text).

12-8 Because this part is constantly holding water, trap bends have carefully fitted swivel joints connecting the J-bend and waste arm. Although no gaskets or washers are used here, plumber's putty or silicone rubber may be applied if the trap is metal, polypropylene, or PVC plastic. A slip jamnut holds the parts together leak-free.

12-9, 10 A P-trap (left) connects its sink with a wall waste stubout. Water is trapped in the J-bend section to seal out gases and vermin, yet lets wastes through. Floor waste outlets are rare, but if your kitchen sink has one, a Genova tubular S-trap (right) will fit it. An extra-deep trap seal prevents siphoning of trap water by the out-rushing flow. While most codes do not permit the use of S-traps in new construction, the one by Genova works well in replacement applications.

12-8

12-9

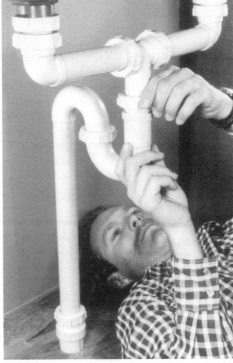

12-10

waste opening misalignments without leaking. P-traps are used with wall waste openings. Genova's P-trap is Part No. 175151 or 175141. (Fig. 12-9).

Another kind of trap, called an S-trap because of its "S" shape, is used with a floor waste opening. It's the only practical way to serve a fixture with a floor waste connection. Trouble is, S-traps are banned by most of today's plumbing codes because they siphon easier than P-traps, sometimes losing their protective water seals. How to convert a floor drain so that a P-trap can be used is shown in Fig. 14-24. It uses a NovaVent™ Automatic Vent Valve. However in most replacement installations, Genova's S-trap (Part No. 177141 or 177151) works fine. (See Fig. 12-10.)

Tubular parts (Fig. 12-11) for kitchen sinks and laundry tubs are 1½" in diameter. Modern house DWV waste openings are 1½", too, so they're both the same size.

The least known and hardest to find

TUBULAR PRODUCTS

Flanged Waste Tailpiece
Dynamic-Duo Trap Adapter
S-Trap
Baffle Tee
Dishwasher Tailpiece
End Opening Continuous Waste
Adjustable Tailpiece
Bathtub Trip Lever Waste
P-Trap
Assorted Slip Washers
45° Slip Coupler
Center Opening Continuous Waste
Tee Slip Coupler
90° Slip Coupler
Slip Coupler
Slip Jamnut
Flanged Tailpiece

12-11

tubular parts are the small slip coupling fittings including 90° and 45° elbows, tees, and straight couplings. To help you plumb, Genova offers them all. You'd likely need these only if you were designing your own under-the-counter tubular system for a double bowl sink.

Continuous waste. You can buy a packaged drain system that includes all the tubular fittings needed. These are available primarily for double-bowl kitchen sinks. You need one type for a sink that's centered on its

waste opening, a different kind for one with a waste opening centered on one of the bowls. Either way, the tubular drain setup is called a **continuous waste.** An end opening continuous waste (Part No. 186001 Fig. 12-12) is used where one bowl centers on the waste opening. A center opening continuous waste (Part No. 186101—Fig. 12-13) is used where the waste opening is between the bowls.

Baffle tee. A fitting that is always part of an end-opening continuous

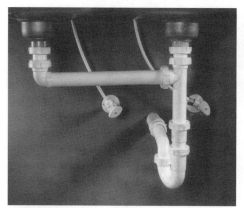

12-12

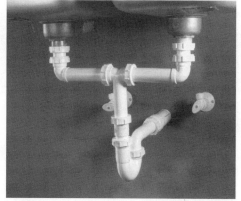

12-13

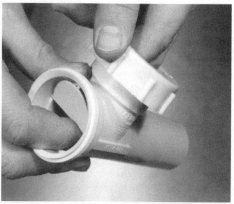

12-14

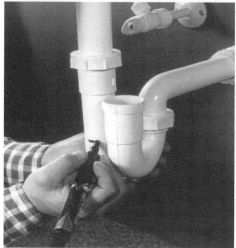

12-15

12-16

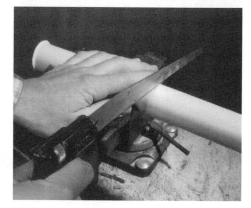

12-17

12-12,13 An end opening continuous waste (left) can be arranged to fit either a right or left-handed double-bowl sink waste connection.

12-14 A deflector plate inside a baffle tee routes disposer wastes toward the trap rather than having them back up into the unused sink bowl. This avoids having to build separate DWV wastes for the sink and disposer. The plastic baffle may be taken out if there is no disposer.

12-15 If a trap is too low to reach the sink's tailpiece, a Genova adjustable extension tailpiece will let it do so. Mark the extension for cutoff.

12-16 Sink strainers come with flat washers that should be installed between the bottoms of the basket drains and the flanged waste tailpieces. These are soft and serve as effective leak-preventers.

12-17 Genova Polypropylene drainage tubes and extensions may be cut with any fine tooth saw. It helps to hold them in a vise while cutting. Make sure the pipe end is free of any burrs.

waste package is a tee. It goes beneath the bowl on the waste connection side. Genova's Part No. 186301 baffle tee contains an inner Polypropylene deflector plate that keeps disposer wastes from backing up in the unused sink drain. The baffle (Fig. 12-14) is easily removed so it can be slipped out if no food waste disposer is being used.

Adjustable tailpiece. Straight sink drain tubes come two ways: as adjustable tailpieces and as flanged tailpieces. Adjustable tailpieces contain slip fittings, and are used to extend drainage tubes.

If your trap is a little short of reaching up to the sink, an adjustable tailpiece is what you need. (See Fig. 12-15.) They're also needed for connection to some food waste disposers.

Flanged tailpiece. These are simple tubes with a flat face on one end. They're used to connect directly to the sink drain basket (and to some food waste disposers).

Both 1½" diameter flanges and adjustable tailpieces come in 6" and 12" lengths, and a 1¼" adjustable tailpiece comes in 6" lengths.

Flanged waste tailpiece. A short tube called a flanged waste tailpiece

(Fig. 12-2) is used to connect a continuous waste up to its sink drains. Two of these come as part of a continuous-waste package.

As with most plumbing products, it's best to stick to what's made by the same manufacturer. That way, you know everything will work together.

Besides these fittings, you can buy extra slip jamnuts (Part No. 147151) and slip washers (Part No. 148001—contains assorted types and sizes, including flat washers for use with sink drain baskets).

Connecting sink drains

When plumbing the drains for a sink, always start at the sink's drain baskets. Basket strainers are threaded on the bottom of the sink to accept slip jamnuts. The nuts go onto 1½" flanged waste tailpieces and hold them to the strainers with flat washers between (Fig. 12-16). Some double bowl continuous waste packages, like Genova's, come ready to connect to the sink. Others omit the waste tailpieces, so you must buy them, also.

When fitting up a double-bowl sink's continuous waste, the horizontal tube (or tubes) that connects between *continued on page 87*

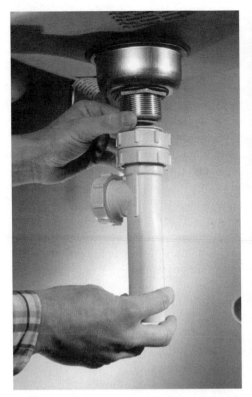

12-18

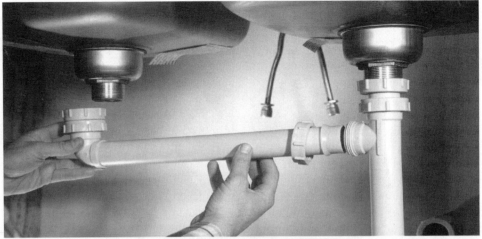

12-19

12-20

12-18 If it's a double-bowl sink with an end opening continuous waste, fit up the baffle tee first. The baffle tee fastens to a short flanged waste tailpiece, which gets a flat washer between it and the sink strainer.

12-19 The tube between the sink bowls of an end opening continuous waste may need cutting to fit less-than-standard bowl centers.

12-20 The end opening continuous waste is attached to the basket strainer at one end and the baffle tee at the other. Tightening the slip jamnuts does the coupling.

12-21 A center opening continuous waste connects similarly except that its tee centers between the two sink bowls. As with an end opening continuous waste, use a flat washer between the flanged waste tailpieces and basket strainers.

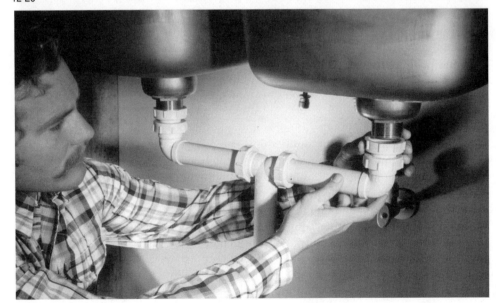

12-21

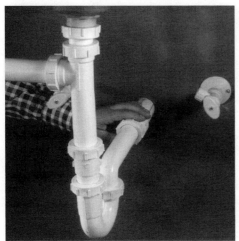

12-22

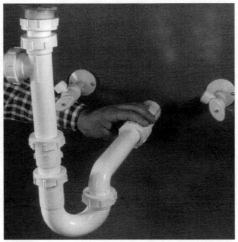

12-23

12-22 A trap swivels to reach a waste opening that is offset somewhat from the sink's drain. At the left, with drain tailpiece and waste opening aligned, the trap aims straight back into its trap adapter.

12-23 To the right, the trap can be swiveled to reach several inches to one side and serve an off-center waste opening.

the two bowls is usually furnished long enough to suit standard-spread bowls. If your sink's bowls are closer together than standard, you'll have to trim the tube(s) to fit, (Fig 12-17) with any fine-toothed saw. Then fit up all the parts, starting with the baffle tee. Fig. 12-19 shows fitting up an end opening continuous waste. Be sure to get the baffle tee on the proper side (Fig. 12-20) to connect to the sink's trap. Fig. 12-21 shows fitting a center opening continuous waste which has no baffle tee and does not need one. The hookup for an end opening or center opening continuous waste is similar.

Disposer. If your kitchen sink has been provided with two separate DWV waste openings, you won't need a continuous waste setup. Instead, you'll need two flanged tailpieces and a pair of P-traps—one for each bowl. Doubled waste openings are intended to make it easy to hook up a food waste disposer. Still, it isn't difficult with just one waste opening, as long as a baffle tee is used to aim disposer discharges toward the trap. Speaking of that, if a food waste disposer is to be added along with new tubular drains, install it before you connect the drains. For how to install one, see Chapter 14.

Trap-connecting

The baffle tee and tailpieces, plus any extensions, should be long enough to connect the trap, leaving its arm inclined slightly in the direction of flow. If they're too long, the tailpieces may be cut shorter. Traps can be swiveled one way or the other to reach an off center waste opening (Figs. 12-22 and 12-23). When working with an S-trap, try to make your length cuts at the upper end to keep the trap's bend up high out of the way of cabinet shelving.

Here's the proper way to install a trap. First, slide a slip jamnut (threads facing downward) up onto the fixture tailpiece or the baffle tee. Follow this with a slip washer—its flat side up (a slip washer's flat face should always aim toward the slip jamnut). The washer will grip the tailpiece, holding the nut in place. Now, push the trap's J-bend up to meet the nut and run the threads up just tight enough to keep everything from slipping off. Slide another slip jamnut, threads first, onto the trap's arm, followed by still another nut, threads last. Slide a slip washer onto the trap arm, flat face first. If you can, use the pro plumber's leak preventive trick shown in Fig. 12-7a & b.

Now you can slide the trap's arm into the trap adapter at its waste opening. If the arm doesn't quite reach, it may be extended with an adjustable tailpiece. If it's too long, you can cut it shorter. Align the two parts of the trap so they meet at the center swivel joint, and start the threads there (Fig. 12-24). Then thread the slip jamnut onto the trap adapter. Don't tighten anything yet, though.

At this point, you're ready to slide the trap's J-bend up or down on the fixture tailpiece or baffle tee to get the desired slight slope in the trap arm. Align everything while loose (Fig. 12-25), then tighten that slip nut. Then tighten the one at the trap adapter (Fig. 12-26). Finally tighten the slip jamnut at the center joint.

Where local codes don't permit slip coupled traps, you can use a solvent welded Los Angeles Pattern P-trap. Figs. 12-27 and 12-28 show a comparison between a slip coupled tubular P-trap and a solvent welded trap.

12-26

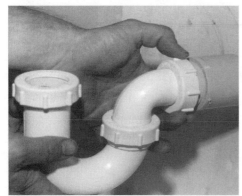

12-27

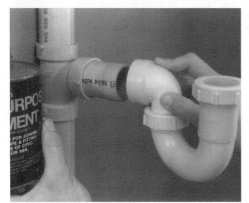

12-28

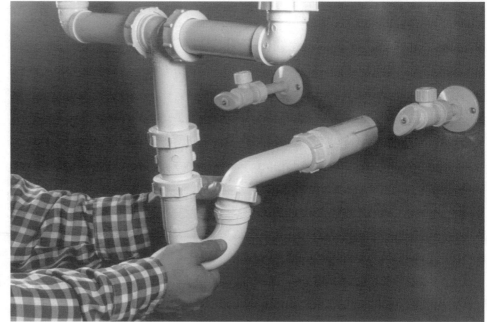

12-24

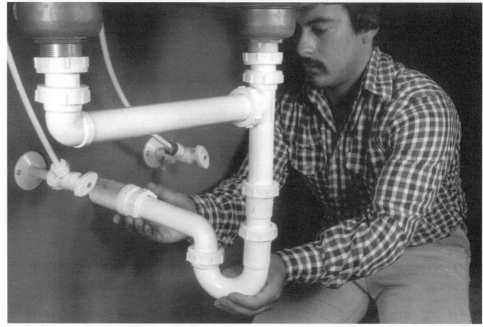

12-25

12-24 To install a trap, get both of its parts into position before making up the center joint. Don't tighten any parts until after all are in place.

12-25 Then, when the entire trap is aligned with a slight slope toward the wall, the fittings may be hand-tightened.

12-26 The last slip jam nut to be tightened is the one on the center joint. If there are leaks when the bowl is test-drained, tighten all the nuts a little more with water pump pliers.

12-27 A tubular P-trap (Part No. 175151) fastens into the DWV system with a trap adapter and slip jamnut.

12-28 A Los Angeles pattern trap (Part No. 78715) joins directly to the waste pipe by solvent welding. This special trap is designed to eliminate slip-connections to a fixture's waste pipe. A few local codes require it.

Section 4 Doing Plumbing Projects
Chapter 13 Plumbing Replacements

Table 13-A

Solvent Welding Emergency Plumbing Replacement Kits

Pipe Type (house)	Pipe Size	Parts	Genova Part Number	Fits Pipe Removed	Tools Needed
Threaded galvanized Steel or brass	1/2"	½" x 10' CPVC tube* 2 ½" CPVC couplings 2 ½" CPVC transition unions Solvent welding kit	50005 50105 530451 141001	4¾" to 10' 4"	Pipe wrench hacksaw pliers or open-end wrench
Threaded galvanized Steel or brass	3/4"	¾" x 10' CPVC tube* 2 ¾" CPVC couplings 2 ¾" CPVC transition unions	50007 50107 530501	5½" to 10' 4¾"	Pipe wrench, hacksaw pliers or open-end wrench
Sweat-type copper Solvent welding CPVC	1/2"	½" x 10' CPVC tube* 2 ½" Genogrip adapters 4 ½" CPVC couplings Solvent welding kit	50005 530751 50105 141001	1³/₈" to 1½" 2⁵/₈" to 10' 1½"	Hacksaw or tubing cutter
Sweat-type copper Solvent welding CPVC	3/4"	¾" x 10' CPVC tube* 2 ¾" Genogrip adapters 4 ¾" CPVC couplings	50007 530851 50107	1⁷/₈" to 2" 3½" to 10' 2"	Hacksaw or tubing cutter

* Cut to convenient storage length-- couplings included in table to suit 40" cut lengths

Your knowledge of do-it-yourself plumbing can be used beautifully in handling plumbing replacements around the house. These consist of projects such as fixing leaks; installing new traps and drains; and replacing an old water heater, toilet, or sink. You stand to save lots of money by doing your own plumbing replacements.

Emergency pipe replacement
If the water in a frost-split water supply pipe is frozen, it won't leak—until it thaws. The emergency can catch you off guard, so it's best to get ready ahead of time. You can do this by putting together a handy repair kit of tubing and the necessary adapters to fit it to your house water supply piping. The kit can be based on the Genova CPVC Solvent Welding System or Universal Fittings. Make one emergency kit for 1/2" pipe and one for 3/4". Tape or wire the parts together and set them aside. Then if you need your emergency repair kit, just quickly cut out the damaged old piping parts and replace with the new ones. The emergency kits are for pipe repair only. If a fitting is damaged, you'll need a replacement for it, as well. The replacement should last as long as your house plumbing.

CPVC kit. If full length 10' solvent welding CPVC tubes are too long for convenient storage, you can section them into shorter pieces, adding one coupling for each saw cut you make. The table shows the minimum and maximum lengths of CPVC repair that can be handled. By adding more CPVC tubing and the necessary couplings, longer replacements can be done. In copper/CPVC, there's a short range of lengths that cannot be fitted, because they fall between what the parts can accommodate. Make your bad-part cutouts short or long enough to avoid these "blind spots." They're indicated in the table.

Universal Fitting kit. A Universal Fitting Emergency Repair Kit can be as simple as a single coupling. Two couplings of each size permit bridging over a longer frost-split tube with a length of CPVC tube between. For threaded piping, use Universal Female Threaded Adapters, plus a length of tubing between them. Table 13-B shows the parts and repair lengths that can be accommodated.

Figs. 13-1 to 13-3 show how to install the tubing-replacement kit. Figs. 13-4 to 13-8 show how to install the threaded metal pipe replacement kit.

Once installed, the thermoplastic tubing insulates against the "battery" effect that metal piping systems develop. On the other hand, if the portion of metal piping that you

13-1

13-1 A close-quarter tubing cutter is useful in cutting out a leaking section of water supply tubing.

Table 13-B
Genova Universal Fitting Emergency Plumbing Replacement Kit

Pipe Type (house)	Pipe Size	Parts	Genova Part Number	Fits Pipe Removed	Tools Needed
Threaded galvanized steel or brass	1/2"	½" x 5' CPVC tube 2 ½" Female adapters 2 ½" Brass short nipples	52421 543051 Hardware	6" to 5' 4"*	pipe wrench hacksaw open-end wrench
Threaded galvanized steel or brass	3/4"	¾" x 5' CPVC tube 2 ¾" Female adapters 2 ¾" Brass short nipples	52431 543071 Hardware	6½" to 5' 4½"*	pipe wrench hacksaw open-end wrench
Sweat-type copper Solvent-welding CPVC	1/2"	½" x 5' CPVC tube 2 ½" Universal Couplings	52421 541051	⅛" to ¾" 2¼" to 5' 2"*	hacksaw or tubing cutter
Sweat-type copper Solvent-welding CPVC	3/4"	¾" x 5' CPVC tube 2 ¾" Universal Couplings	52431 541071	⅛" to ¾" 2¼" to 5' 2"*	hacksaw or tubing cutter

* Greater with longer CPVC tube

13-2 To work with Universal Fittings, slide a coupling from your kit onto the chamfered and lubricated tube ends. You must be able to push the copper tubes apart about 3/4".

13-3 Make sure that both tubes enter the fitting fully. Finally, tighten the hand collars on the coupling, and you've made a leak-free repair.

13-4 To replace a leaking threaded metal pipe with your emergency kit, simply saw through the defective pipe. This one had been split by freezing. The kit will replace pipes that leak for any reason.

13-2

13-3

13-4

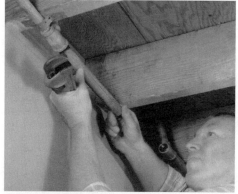

13-5

13-6

13-7

13-8

13-9

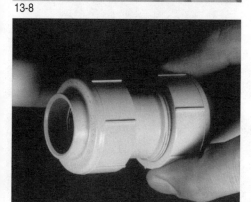

13-10

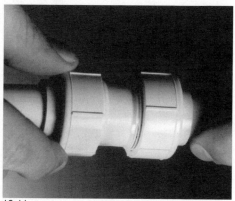

13-11

13-5 Next, unthread both ends of the leaking pipes with a pipe wrench. You need enough space between fittings to install the essential kit parts: when using CPVC, the two transition unions and a short linking CPVC tube between them.

13-6 Now thread in transition unions from the right-sized emergency kit, using one at each end of the old metal pipe. A pair of water pump pliers or an open-end wrench fits between the CPVC collar and metal fitting. Use pipe dope or TFE tape on the male threads.

13-7 The last step is to install the solvent welded CPVC repair. To make it, measure between the two transition union gaskets and deduct 1" for makeup. Cut the CPVC replacement tube that length, solvent welding it to the transition unions.

13-8 Here's how the replacement CPVC fits in. The transition union collars clamp their elastomeric gaskets tightly to join plastic and metal without leaks.

13-9 To fix a leak in a CPVC or copper tube, turn off the water and cut through the tube at the leak.

13-10 Separate the tubes enough to slide a Genova Universal Coupling of the correct size onto one side.

13-11 Slide the tube on the other side into the coupling. Then tightening the hand-nuts at each end completes the job. The repaired pipe is ready for immediate use.

replaced with thermoplastic tubing is part of an electrical grounding system, to meet National Electrical Code (NEC) requirements, ground clamps and a bonding wire must be installed across the plastic section. Have this done by an electrician.

Patching

If your solvent welded CPVC or PVC pipes should ever leak, you can put quick, lasting solvent welded patches right over the leaks. Because thermoplastic piping does not corrode, the most likely cause of a leak is a nail driven through the pipe wall, or a frost-split pipe.

Universal Couplings. One patching method is cutting or sawing the pipe apart at the leak and bridging over it with a Universal Coupling. (See Figs. 13-9 to 13-11.) If it's a lengthy leak, such as a split pipe, a new length of pipe can be installed using two Universal Couplings.

Repair couplings. On DWV pipes repair couplings may be used. These are ones without inner stops or shoulders that will slide all the way onto a same-sized pipe. They are available for Genova DWV Pipes from 1½" to 6" in diameter, including 3" Schedule 30 pipe.

Outside patch. A hole—small or large—is easily fixed by solvent welding a patch over it. Use a piece cut from a scrap piece of pipe. Patching a hole in a DWV pipe is shown in Figs. 13-12 and 13-13.

Renewing traps and drains

According to a recent survey among do-it-yourself plumbers, replacing traps and drains is the most common

(Continued on page 93)

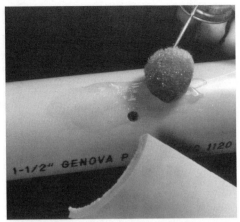

13-12

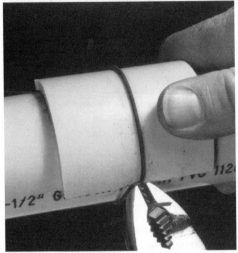

13-13

13-14

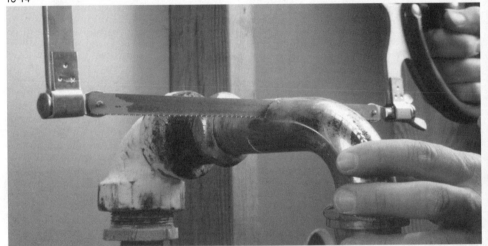

13-15

13-12 To repair a nail or drill hole in a pipe—PVC-DWV or CPVC water supply—cut a patch from a scrap piece of pipe. It should be big enough to extend an inch or so beyond the hole on all sides. Then turn off the water and apply a heavy coat of solvent cement to both the outside of the pipe and the inside of the patch.

13-13 Immediately join the patch centered over the leak. Wire it in place until the solvent cement has had a chance to set. The wire may be removed later.

13-14 Old drains can be removed by loosening the slip jamnuts and slipping the parts off. A large pair of water pump pliers can exert lots of force on a stubborn, corroded nut.

13-15 Last resort is hacksawing the old drains off. You'll still need to get the slip jamnuts at trap adapter and fixture off, but these can be loosened by sawing through them crosswise. Be careful not to saw into the male threads. With the tubular drains off, you'll have more space for working on the nuts.

13-16 To remove a sink from its water supply, it's often simplest to cut off the old copper supply tubes at a convenient working height. Here it's done with a close-quarter tubing cutter. The water supply hookup to the fixture will be replaced from this point using straight supply valves (Part No. 530301). These fit over the copper tubes without solvent welding or soldering.

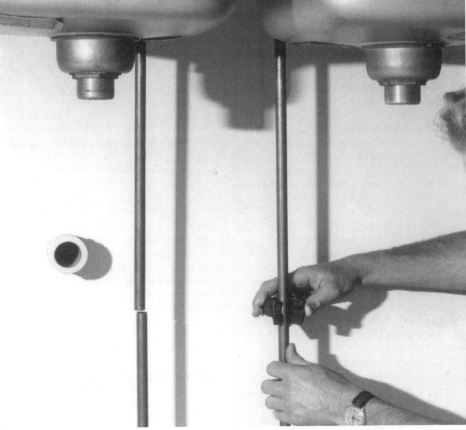

13-16

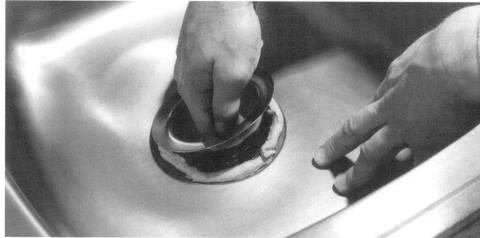

13-17

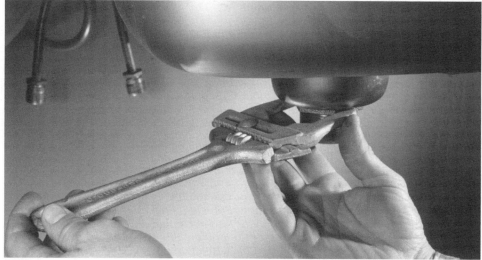

13-18

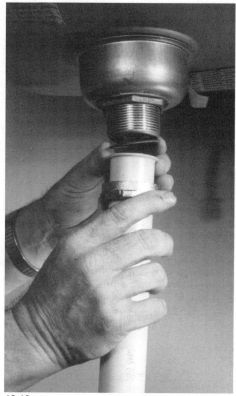

13-19

13-17 A new sink bowl comes without its strainers in place. You install them by placing a ring of plumber's putty around the drain opening and lowering a strainer over it.

13-18 The strainer gets a washer and a nut below the bowl. When the nut is tightened with a fixture wrench or monkey wrench, the strainer is secured in the sink bowl. Then go topside again and remove the excess plumber's putty from around the drain.

13-19 Tailpieces are held to the bottom of the strainer with metal slip jamnuts, also furnished with the sink. If a tailpiece does not come with the sink, you can get a Genova polypropylene one 6" or 12" long. Install it with a flat washer and slip jamnut.

household plumbing project.

Removing the old. Much of the information in the previous chapter on new drain installations also applies to drain replacements. The only difference between renewing and installing new tubular fixture drains is the one additional step of getting the old drain hardware off. It's best to tackle this starting with easy methods and progressing to more difficult ones if necessary. Begin by trying to loosen the old drain's slip jamnuts with a fixture wrench, pipe wrench, or monkey wrench. Water pump pliers will work, too (see Fig. 13-14). This will allow you to disassemble the old drains.

If any of the nuts won't turn, squirt some penetrating oil around the threads and wait a bit for the oil to soak in. Tapping on the nut lightly with a hammer helps the oil to penetrate. Then try removal again. As a last resort, bring on a hacksaw (Fig. 13-15) and cut through the old tubes. Have a pail beneath the trap to catch the water held by it. Naturally,

the water to the fixture should be turned off.

With the old parts off, your way is clear to install the new ones. Genova's Polypropylene Tubular Products with their hand-tightened slip couplings make drain replacement a pure breeze. And they'll outlast the old metal tubular drains and traps.

It's okay to make a partial replacement; that is, replace just the trap, if it's all that's damaged. Unless they show corrosion, fixture tailpieces normally need not be replaced. Be sure, however, to use new slip washers on whatever is being replaced.

Although you may use 1½" tubular parts for smaller 1¼" washbasins, if your wall or floor waste fitting for a washbasin is 1¼", then you should get all 1¼" tubular goods, also. This avoids the horrible practice of going from 1¼" to 1½", then back to 1¼" farther down stream. No plumbing code approves that, and it spells endless problems with clogged drains later.

Some waste pipes are plumbed to come in at an angle or are off center with their fixtures. A P-trap will adjust to fit them. If it's not long enough after being swiveled to one side, an extension tailpiece (6" or 12" or one cut to length) will help it reach. A trap should extend into its waste connection about 1½", but not too far.

If replacing the drains on a two-bowl kitchen sink with separate DWV waste openings for each bowl, you'll use two flanged tailpieces, possibly two adjustable tailpieces, and a pair of traps. In that case, a continuous waste setup won't be needed.

Water heater connections

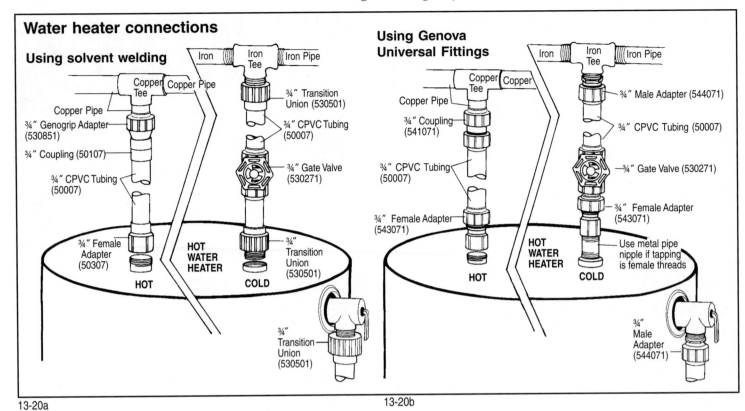

Using solvent welding

Iron — Iron Tee — Iron Pipe

Copper Tee — Copper Pipe

Copper Pipe
¾" Genogrip Adapter (530851)
¾" Coupling (50107)
¾" CPVC Tubing (50007)
¾" Female Adapter (50307)

¾" Transition Union (530501)
¾" CPVC Tubing (50007)
¾" Gate Valve (530271)
¾" Transition Union (530501)

HOT
HOT WATER HEATER
COLD

¾" Transition Union (530501)

Using Genova Universal Fittings

Iron — Iron Tee — Iron Pipe

Copper Tee — Copper

Copper Pipe
¾" Coupling (541071)
¾" CPVC Tubing (50007)
¾" Female Adapter (543071)

¾" Male Adapter (544071)
¾" CPVC Tubing (50007)
¾" Gate Valve (530271)
¾" Female Adapter (543071)
Use metal pipe nipple if tapping is female threads

HOT
HOT WATER HEATER
COLD

¾" Male Adapter (544071)

13-20a   13-20b

New sink/washbasin

Replacing a sink or washbasin goes a step beyond a mere change of tubular drains. Besides getting the old drain hardware off, you'll be removing the fixture's faucet from its water supply. This is simple to do by cutting off the old pipes (Fig. 13-16). Then the faucet, or faucet and sink can be lifted free. Most sinks installed in cabinets are held from below by clamps. Loosening the clamp screws lets the bowl be lifted out of the countertop. Wall mounted sinks are supported by a concealed hanger, and can be lifted off, once the piping is disconnected.

If the water supply pipes serving the fixture were 1/2" (5/8" O.D.) CPVC or copper tubing, you can fit them up directly with angle or straight supply valves. This readies the new fixture for a Poly Riser hookup. If they're 1/2" threaded metal pipes, a pair of Special Female Adapters (Genova Part No. 50305) will ready them for CPVC couplings, 1/2" CPVC tubing stubs, and then fixture supply valves.

Install the new sink or washbasin following directions furnished with it. Plumber's putty is often needed around the outer rim of an in-counter basin as well as around the basket drains (see Figs. 13-17 and 13-18).

Tailpieces. If a tailpiece is furnished with the fixture bowl, use

it. If not, a Genova flanged extension, 6" or 12", may be connected to a sink's drain (Fig. 13-19). If it's a double-bowl sink, a continuous waste system is used.

Change a water heater

A water heater generally lasts about ten years. This makes replacing a heater one of the more common home plumbing projects. Water dripping from the old water heater is an indication of tank failure and is the most typical water heater trouble. Your dealer will help you to select the new heater you need.

If the old tank's drain valve is what leaks, it can be replaced. Simply use a Part No. 530501 Transition Union and a Part no. 530151 Universal Line Valve. Then, a Part No. 53128 Male Hose Adapter solvent welded into the line valve's outlet will allow connection of a garden hose for draining.

The modern way of installing a water heater makes the job go easier and better. It involves the use of CPVC tubing.

Removing the old one. Follow all directions with the new water heater you get. Start your switch-over by turning off the energy to the heater. With an electric heater, cut off the power to the heater. Never depend on a switch, fuse block, or circuit breaker being off. Always test for it with a working neon test light (Fig.

13-21). Most electric water heaters are served by 240 volts, which is double the voltage of your normal receptacle and lighting circuits. You can get killed by touching the live wires inside the heater's junction and thermostat boxes. We recommend having an electrician handle that part of an electric water heater change for you.

If you have a gas water heater, turn off the main gas cock—the one in the pipe leading to the water heater control—and check to see that the pilot light on the gas heater is out. Then disconnect the gas pipe from the water heater control valve. Plug or cap the piping so it can't leak gas into the house.

Next, turn off the cold water inlet valve leading to the heater tank. Drain the heater by attaching a garden hose to the drain valve at the bottom of the heater and letting the water run out of the old heater. Lead it away to a good draining spot. Opening a hot water faucet will help it to drain. Just remember that the water coming out of the heater can be very hot.

While the heater tank drains, you can remove the old water heater's piping. If there are no handy unions, the easiest way is to just cut through the pipes (Fig. 13-22). With a utility room installation, cut through the horizontal sections of pipe rather than the vertical, as it makes for

13-21

13-22

13-23

13-24

13-21 Have an electrician handle the wiring on an electric water heater. Before touching any bare wires inside the heater's junction box, he will turn off the power to the heater. Then, use a neon test light to be sure none of the wires are "live." The light should not light when contacted across every combination of wires and the grounding screw.

13-22 If the old heater's piping has no unions, you can get the old plumbing free by sawing through the pipes, removing threaded stubs from the hot and cold mains. Mark the pipes "hot" and "cold." First, you'll have to turn off the house main shutoff.

13-23 Jockey the new heater into position, level it, and begin the hookup. It helps to "walk" the heater along on its feet or leave it in the carton until placed. If you have one, use a dolly.

13-24 Every flue pipe joint for a gas or oil water heater should be locked together with one sheet metal screw per joint. Drill a small pilot hole to get each started. Use the proper type of vent for the heater.

easier replacement runs.

Disconnect a gas heater's vent pipe from the draft hood removing the sheet metal screw that holds the two together. Discard the old draft hood. Remove any earthquake straps from the wall.

The old heater can be carted away, and the new one brought in (Fig. 13-23). Reinstall the earthquake straps if you live in an area that requires their use. Use the new draft hood that comes with a gas heater, securing it to the vent pipe with one sheet metal screw (Fig. 13-24). Never install a gas heater without a draft hood.

Water Connections. Using Genova threaded metal Transition Unions lets you connect directly to the heater's tappings. Fig. 13-20a shows the details of an installation using solvent weld fittings. If the heater comes with female threaded tappings, fit these with a transition union, as shown on the right. If it comes with male threaded hot and cold water stubs, fit them as shown on the left. At the hot and cold water mains, fit a CPVC or copper tubing main, as shown on the left-hand side; fit a threaded metal main as shown on the right-hand side. No sweat soldering or pipe threading will be needed.

Fig. 13-20b shows the details of a Genova Universal Fittings installation. If the heater comes with male threaded hot and cold water stubs, fit them with Universal Female Adapters, as shown on the left. If the heater comes with female threaded tappings, fit them with metal pipe nipples and use the Universal Female Adapters, as shown on the right.

13-25 Hot water transition unions should be installed on the heater. The male threads are wrapped with TFE tape or treated with pipe dope and then can be screwed in. Tighten with water pump pliers or an adjustable open-end wrench. The use of transition unions is vital to successful water heater hookup.

13-26 Next, install a Genova ¾" Gate Valve (Part No. 530271) over a short stub of ¾" tubing solvent welded into the transition union on the heater's cold water inlet. The valve comes with Genogrip Adapters at both ends for an easy hand-tightened connection.

13-27 The Genova ¾" Gate Valve works like a standard brass gate valve, opening to full bore. For use in ¾" high-flow applications as with water heaters, the valve's straight-through design doesn't impede water flow.

13-28 The body and gate of the Genova valve are brass, just like the best threaded gate valves. Unlike standard gate valves, the valve's Genogrip connections mate with both types of ¾" water supply tubing—CPVC and copper.

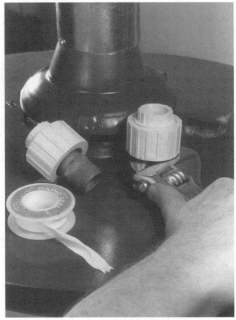

13-25

13-26

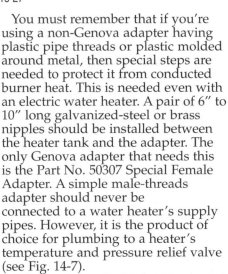

13-27

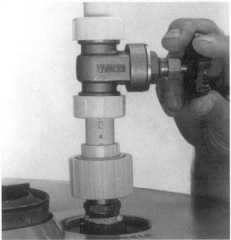

13-28

You must remember that if you're using a non-Genova adapter having plastic pipe threads or plastic molded around metal, then special steps are needed to protect it from conducted burner heat. This is needed even with an electric water heater. A pair of 6" to 10" long galvanized-steel or brass nipples should be installed between the heater tank and the adapter. The only Genova adapter that needs this is the Part No. 50307 Special Female Adapter. A simple male-threads adapter should never be connected to a water heater's supply pipes. However, it is the product of choice for plumbing to a heater's temperature and pressure relief valve (see Fig. 14-7).

CPVC will handle higher-than-rated temperatures on an intermittent basis, so there's no worry about burner heat affecting the piping. However, don't try to hook up a water heater using PVC, PE, or ABS plastic pipes. None

of these can take heat.

The Part No. 530271 ¾" Gate Valve (shown sectioned in Fig. 13-27) goes in the cold water inlet side of the heater (Fig. 13-26). This position will be marked "cold" on the heater. Only a full flow valve should be used, and the valve should be open all the while the water heater is being used. The unique valve is designed for direct connections in CPVC or copper systems.

The ¾" Genova Gate Valve is designed both for controlling water flow in newly installed thermoplastic water supply piping and for replacement use in existing plumbing. It fills a need as part of the Genova line of hot/cold water supply tubes and fittings.

Figs. 13-20 to 13-33 show how to plumb a water heater.

If your existing piping is ½" rather than ¾", use the same kinds of

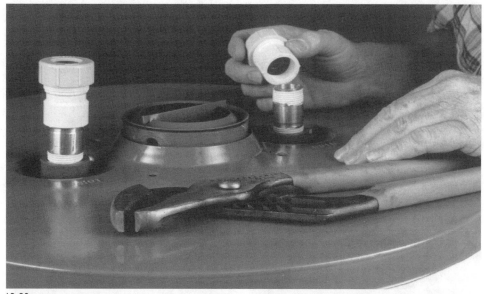

13-29

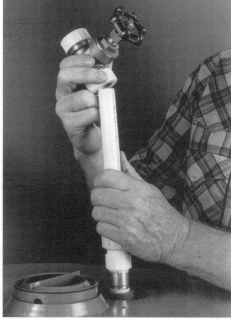

13-30

13-31

13-29 To hook up a water heater without solvent welding using Genova's Universal Fittings, connect to the heater's hot and cold water tappings with ¾" female adapters. If the heater comes with female tappings, install short brass nipples so that female adapters can be connected to the heater. Simple male threads adapters should not be used as transition fittings. Wrap TFE tape on all male threads.

13-30 Then slide a short 3/4" CPVC tube into the female adapters on the cold water side and install a 3/4" gate valve atop it. The connections push in and hand-tighten.

13-31 At the upper end, the CPVC tubes fit copper hot and cold water mains with Genova Universal Couplings. No sweat soldering is necessary.

fittings as shown in Fig. 13-20a or 13-20b, but in the smaller size. Then you can use the Genova Universal Line Valve, Part No. 530151.

Whether the old heater had one or not, every new water heater should be fitted with a new temperature-and-pressure relief valve. This prevents an explosion in case the heater's energy cutoff safety features should fail. You can make the relief system with CPVC tubing and fittings. How to do this is described in the next chapter.

For making the gas connection to your new water heater, remember that CPVC is not suitable for this. Be sure to test all gas connections—including the heater's gas valve—for leaks using a solution of dish detergent and water. With LP-gas, use a pure soap solution, as it works better with LP gas than detergent does. A new electric heater should be electrically grounded using the green hex-headed bonding screw in the junction box. Your electrician should take care of this. We recommend hiring a pro for electrical work.

Changing a toilet

When you modernize a bathroom, you'll likely change the toilet, too. Or perhaps the one that's already there will be removed and put back. Either way, the process isn't complicated.

Removal. Most household toilets are the two piece type with separate tank and bowl. To take one out, turn off the water and flush the toilet, holding the trip lever down until all water runs out of the tank. Sponge

13-32 Cut off the heater's cold water supply tube long enough to slip into the ¾" gate valve and make up that connection.

13-33 Complete the hot water side without a valve by running a ¾" tube directly to the heater's female adapter from the hot water main. Tightening the Genova Universal fittings by hand makes them leak-free.

13-34 A wax toilet gasket installs around the outlet horn of the toilet bowl to seal the connection between bowl and DWV system. The flat side goes against the bowl. The wax gasket, available widely, is easiest to use for this. To seal between bowl flange and floor, you may want to stick on a slim ring of plumber's putty.

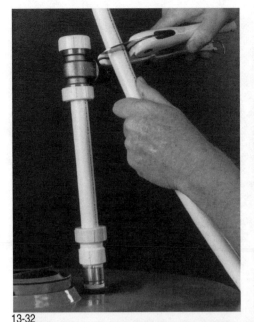

13-32

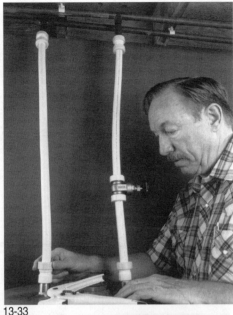

13-33

13-34

out the water that's left in the tank and bowl. Next, disconnect the tank from its water supply at the wall or floor.

If the tank mounts to the wall, remove the two slip jamnuts on the large tube between tank and bowl, separating them. Now the tank can be taken off the wall by removing its bolts.

If the tank mounts atop the bowl, look inside on its bottom for two large screws with rubber washers under them. Taking these out will allow the tank to be lifted from the bowl. You may have to use pene-trating oil on the nuts below the bowl's flange, and hold them with pliers or locking pliers while turning

the screws from inside the tank. Once it's off, place the tank where it won't get broken.

The bowl is held to the floor by two (sometimes four) bolts with nuts. They're usually hidden underneath rounded caps at the base of the bowl. Plumber's putty or a snap-on connection secures the caps, and they'll pry off, exposing the nuts for removal. Penetrating oil may help loosen the nuts. Saw them off with a hacksaw if you must, placing masking tape strategically to protect the bowl from scratching.

Once the nuts are off, the toilet bowl will lift free from the floor and the connection with its soil pipe. Tilt the bowl forward to keep from

spilling and have a pail handy to catch water held by the built in trap. Set the bowl upside down on a padding of newspapers.

Removing the bowl exposes the end of the soil pipe and toilet flange. Stuffing a cloth into the pipe end will keep sewer gases out of the house. The old putty or wax gasket ring between bowl and toilet can be scraped off with a putty knife, leaving the toilet flange clean. Put a new pair of brass toilet hold-down bolts on the toilet flange.

Install a toilet. To put the old toilet back in or to install a new one, start with the bowl upside down. Follow specific instructions with the toilet you buy. Use a wax bowl gasket

13-35

13-36

13-35 Lower the prepared bowl squarely onto the toilet flange so the mounting bolts enter their holes in the bowl's base.

13-36 Pressing down, rock and slightly twist the bowl. This squeezes down the wax gasket until the bowl reaches the floor square with the wall.

13-37 A toilet bowl is fastened down with a pair of bolts attached to the toilet at the floor flange. Tighten a turn at a time each. You want them snug, but not so tight that the bowl flange is broken. Recheck them the next day. Color-matched caps filled with plumber's putty or snapped on will be placed later to cover the sawed-off bolt ends.

13-37

(standard hardware item) around the bowl's outlet horn (Fig. 13-34). Unless the soil pipe is 4", the wax gasket should be the kind without a sleeve. Either plan to caulk between the bowl and floor later or put a ring of plumber's putty around the bowl's base flange where it will meet the floor. Then invert the bowl, placing it over the flange as shown in Figs. 13-35 and 13-36. Fasten it down (Fig. 13-37) being careful not to overtighten the brass bolts. (Remember that china breaks!)

Next, mount the toilet tank to the bowl or wall. (If it's an integral tank/bowl toilet, this step isn't necessary.) The soft donut-like spud washer that goes between tank and bowl should be placed on the tank, flat face toward the tank. A little ring of plumber's putty on the spud washer is the professional plumber's secret to no leaks. Set the tank on the bowl and fasten the tank to its bowl with two hold-down bolts and washers. The rubber washers should go against the inside of the tank to prevent leaks. Plumbers putty should be used here, too. Be sure to use washers on the lower end, also, to protect the bowl from breakage. Tighten the bolts a little at a time as they should be snug enough to prevent leaks, but not so tight as to strain and possibly break the tank or bowl. The tank should end up level.

It may be best to put in a new toilet water supply valve along with the change. If it's metal, the old one is likely corroded. Use a Part No. 530651 Angle Supply Valve for a wall supply and 530301 Straight Supply Valve for a floor supply. A Closet Poly Riser (Part No. 530801 (12") or 530821 (20")) will let you join the toilet tank to its water supply (Fig. 13-38). Before leaving the project, turn on the toilet and test-flush it a few times to check for leaks.

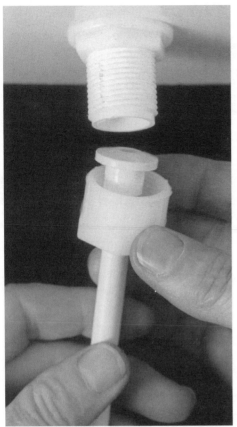

13-38

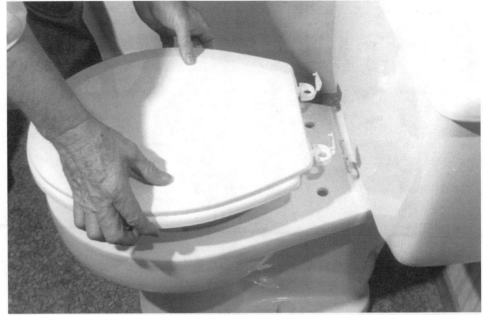

13-39

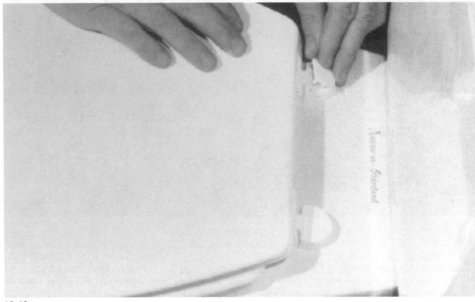

13-40

13-38 Connect the toilet tank to its water supply with a toilet Poly Riser and the tank nut. This nut is usually furnished with the toilet. Light wrench-tightening should make it leak-free.

13-39 To install a new toilet seat, lower the seat into place on top of the toilet bowl, getting the mounting holes aligned.

13-40 After installing and tightening the noncorroding mounting bolts, snap the integral covers down over the bolt heads to conceal them.

Chapter 14 New Plumbing Installations

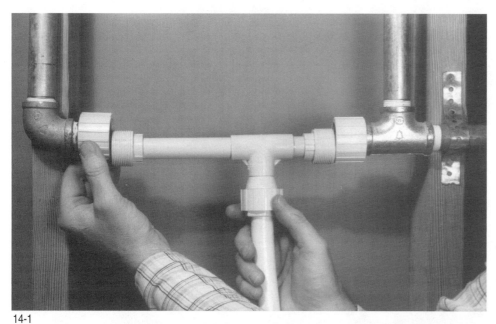

14-1

14-2

14-3

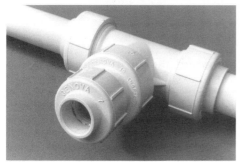

14-4

14-5

14-1 Here's how to put a full flow hose water supply tap into a threaded metal cold water main. It's done by removing a section of the metal pipe, installing two transition unions, then replacing the old section with a new solvent welded CPVC run containing a CPVC tee. Then the addition of a Genogrip Adapter permits the branch run to be made without solvent welding.

14-2 Genova Universal Tees installed in hot and cold copper water supply mains allow easy tapoffs to be made. Branch taps may be in CPVC simply by pushing the tube ends into the fitting all the way and tightening the hand collars.

14-3 By installing a Reducing Assembly in the 3/4" Genova Universal Tee branch, a 1/2" tube fits the tee directly. Ready-made 3/4" x 1/2" reducing tees are also available.

14-4 A 3/4" Genova Universal Tee with a Reducing Assembly installed is ready to accept 1/2" tubing.

14-5 By using the Reducing Assembly, branches can be a size smaller than the mains.

Your home might benefit from additional plumbed-in features. Here's how to plumb lots of them, using easy-does-it Genova thermoplastic piping products.

How to tap into pipes
Whenever you add to your plumbing, you'll need to tap in to existing water supply pipes by installing tees in the lines.

Threaded metal supply piping can be tapped by turning off the water, cutting the pipe, and unthreading the pieces from their fittings. Male threaded Universal Adapters (cold water only) or male MIP transition unions (hot water) are threaded in where the pipe stubs came out. A length of tubing and a tee replace the piping removed. This is shown in Fig. 14-1.

To install tees in water supply tubing—whether plastic or copper— use Genova Universal Tees of the proper size. Turn off the water and cut out a short piece of the old tubing. In its place, install the tee. The tee branch lets you take off with your new run of piping. Figs. 14-2 to 14-5 show branching a new run of piping from copper water supply mains.

Adding a hose bibb
Having a sillcock where you need one can save lots of garden hose work. Start by making two "where" decisions: where to put the outlet and where to get the water for it.

Genova markets a sillcock for nonfreezing climates (Part No. 530941). It is supplied via a 1/2" CPVC tube from inside the house taken from a 3/4" or 1/2" cold water pipe. Tubing runs may be in a basement or crawlspace. In a slab type house, the tubing may be run from an attic, down inside a wall to sillcock height. Because of the lack of

14-6 Genova's Solvent Welding Sillcock attaches to the house wall and fits a garden hose. It accepts a 3/4″ CPVC tube directly or a 1/2″ CPVC tube with a 3/4″ x 1/2″ Reducing Bushing.

14-6

access to make such a tubing run, the latter routing is much harder than the other two.

Where to tap. You must tap into a water line for a hose bibb before it reaches the water softener, but don't tap any branch pipe serving a shower or any fixtures with restricted flow problems. The tapping should be a full flow one. Do not use a saddle tee.

Make your run to the sillcock in 1/2″ CPVC tubing.

The regular Genova sillcock solvent welds to the end of a 3/4″ CPVC tube. It also solvent welds to a 1/2″ CPVC tube using a Part No. 50275 reducing bushing.

If you want to be able to drain the water supply tubing run from inside the house to keep it from freezing, put a Part No. 530161 Universal Line Valve with waste inside the house, either at the sillcock or at the tapoff tee.

To install the sillcock, bore through the house wall and fasten the sillcock to the wall with two rust-resistant screws (Fig. 14-6).

Relief system

Every water heater should have a combination temperature & pressure relief valve, plus a CPVC pressure relief line leading away from it.

Check your own water heater right now. If it lacks these safety needs, waste no time in installing them. The relief valve conducts escaping hot water to a safe discharge point.

Temperature & Pressure (T&P) valve. The T&P valve you get should offer dual pressure and temperature protection. This means that two ratings should appear on the valve's body or identification plate, one for pressure, another for temperature. The valve's ratings should meet local codes. A typical T&P valve for domestic use has a pressure relief setting of 125 psi, a temperature relief setting of 210° F, and a BTU rating of 180,000 to 200,000. It usually contains 3/4″ threads on both inlet and outlet. The T&P valve threads into its own tapped opening high on the side or on top of the heater tank. If used on a water heater with a top tapping, its temperature sensing element should be long enough to reach into the water within the top 6″ of the tank.

Relief line. The second important need is for a pressure relief line to carry away water released by the T&P valve. Some plumbing codes require that the relief line be made from rigid pipe. CPVC is the easiest working piping material meeting these requirements. Rules for relief lines are simple: the tube should be the same size as the relief valve's outlet—usually 3/4″. The line must run full size for its entire length. No valves or other restrictions should be located between the T&P valve and tube end, and the relief line should slope slightly toward its drainage end, but with no low spots to trap water. The end of the tube must not be threaded. Neither should it contain a fitting.

Since there is no continuous water pressure on the joint between T&P valve and relief line, a simple, low cost Genova 3/4″ Male Adapter may be used (Part No. 50407). Thread it into the valve's outlet using TFE tape or pipe dope on the male threads. Tighten it one turn beyond hand tight, but not so tight that you strain the fitting. The 3/4″ relief tube will fit the male adapter (Fig. 14-7). Proceed with CPVC tubes and fittings as needed, all the way to the end.

Keep the relief line as short and direct as possible. It may drain outdoors next to the house foundation (Fig. 14-8) or may empty into a floor drain. It need not "look" directly into the drain, however. It can be placed so that water can flow to the drain. You must allow 6″ between the end of the relief line and the floor.

If there is no other way and your code permits, the relief line may drain into a house waste pipe. Then you must leave a 1″ air gap between the end of the tube and the overflow into which it discharges. This is to avoid creating a cross connection.

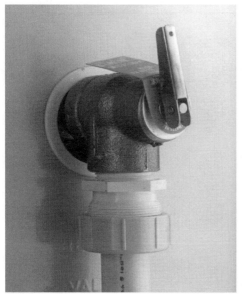

14-7

14-8

14-9

Water softener

It's rare that a water supply cannot be improved by the addition of a water softener. You can install it yourself, using CPVC tubing.

Where to tap. Softener location is all-important. Where you hook into your house plumbing determines what water will be softened and what will not. All or little water can be made soft. Water to outdoor sillcocks should not be softened. If it can be worked, water to toilets should be hard, too. Sometimes just hot water is softened. This is simple. Merely plumb the softener into the water heater's cold water supply. But to get the full benefit of soft water, both hot and cold water should be softened.

When locating the softener, also remember that sacks of softener salt may have to be carried to it and salty regeneration wastes drained away. If it has an electric timer, you'll need a receptacle nearby.

Since a regenerating softener needs a certain amount of water flow to effectively recharge it, be sure your water system can provide it. Something like 3 gallons a minute at 20 psi pressure usually does it.

Softener plumbing. Follow the manufacturer's instructions in plumbing your softener. For the normal solvent welding installation, a hookup needs three Genova Gate Valves (Part No. 530271) or three Genova Universal Line Valves (Part No. 530151). The gate valves give the fullest flow. Be sure that the flow arrows on all three valves point in the direction flow.

With the two lower valves turned off and the bypass valve turned on, the softener is short circuited for replacement or repair. In use, the two lower valves are opened and the bypass is closed. This way, water flowing through the main must pass through the softener before continuing its journey. Thus, it gets softened.

To install the valves, remove a section of the 3/4" cold water main, installing 3/4" CPVC tees (Part No. 51407). The lower valves will solvent weld directly into the tee branches. If you plan it right, the bypass valve will solvent weld into the tee runs with just two 3/4" CPVC couplings (Part No. 50107). From there, you make the inlet and outlet runs to the softener (Fig. 14-9). Adapt at the softener with transition unions—MIP or FIP, depending on what's there. These offer easy take-apart for quick softener removal. The addition of 3/4" Genogrip Adapters (Part No. 530851) lets the connections be made without solvent welding.

Fig. 14-10a shows a typical solvent welded softener hookup; Fig. 14-10b shows a typical Genova Universal Fitting hookup. While it would not happen this way, to illustrate both kinds of hookups, the outlet main on the left is shown in threaded iron metal, the inlet main on the right in copper tubing. Your house will have one type of pipe or the other, but not likely both. Gate valves are used for greatest flow.

14-7 Every water heater needs a relief system. Start a relief line at the Temperature and Pressure relief valve with a 3/4" MIP adapter. Join the first length of tubing into it. Aimed down, as this Universal Fitting one is, the relief line will end 6" above the floor, emptying into a floor drain.

14-8 Or the relief line may exit outside the house and be elbowed down. The end of the CPVC drop tube should have no fittings or threads, it should aim down, and it should end 6" to 24" above grade.

14-9 The 3/4" CPVC tubes are cut to length and adapted to the softener with take-apart transition unions. This softener was designed for direct use of CPVC tubing by solvent welding. Most softeners would need transitions.

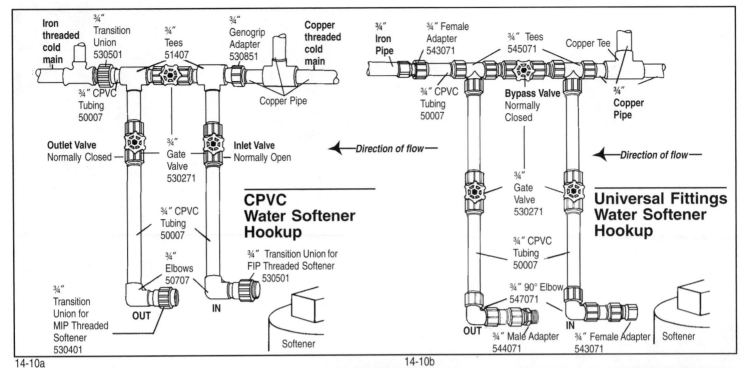

14-10a   14-10b

14-11 To mount a large water filter, with or without a bracket, start by creating a solid frame for mounting it. Here a 2″ x 4″ was notched at the ends and nailed between joists. It serves as a header positioned correctly to align the filter with its piping.

14-12 Now you can turn off the water and cut the cold water pipe. Remove enough to accommodate the filter, allowing ample tubing for the inlet and outlet fittings. While many filters come without any means of mounting, quite a few are supplied with adapters for plastic and copper tubing. Use these, if suited. Here rubber collars fit around the CPVC tube ends. Flange nuts draw them up leak-free.

14-13 Worm drive hose clamps make ideal mounts for all sizes of round filters. If you don't have any clamps long enough, you can join two or more by unthreading the ends and threading them back into a larger clamp, as has been done here. Punch your mounting hole in the clamp band instead of drilling.

14-14 Finally, install the filter and its plumbing, giving a final tightening to make sure it won't rotate in its mount.

14-15 If you know that a water filter is securely fastened to the framing, you needn't be afraid to tackle canister removal even with a strap wrench. The mount, not the piping, absorbs the torque.

14-16 Small under-cabinet filters may come with mounting brackets that you can fasten securely to the cabinet wall. Sometimes, screws for attaching them to metal cabinets are provided. For wood cabinets, wood screws may be substituted. Then when you take off the filter canister, the bracket takes the torque. Here the flexible Poly Risers connected with 3/8″ compression adapters make an easy filter hookup between the fixture supply valve and cold water faucet.

14-11

14-12

14-13

14-14

14-15

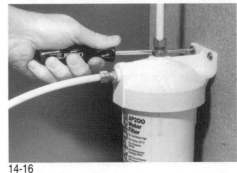

14-16

Wall mounted filter installed with elbows

Cut out section of pipe, install wing elbow and fasten to wall. Mount filter to wall and run pipes to it.

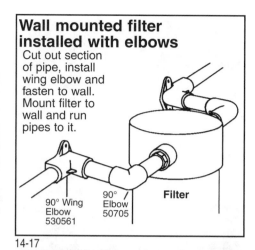

90° Wing Elbow 530561
90° Elbow 50705
Filter

14-17

Cut-away showing how a water hammer muffler works.

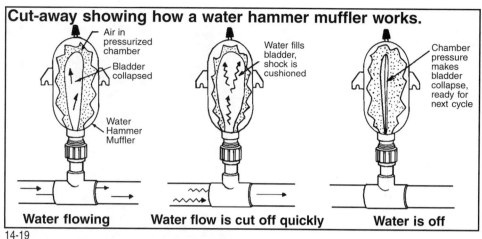

Air in pressurized chamber
Bladder collapsed
Water Hammer Muffler

Water fills bladder, shock is cushioned

Chamber pressure makes bladder collapse, ready for next cycle

Water flowing   **Water flow is cut off quickly**   **Water is off**

14-19

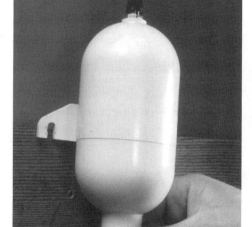

14-18

14-20

14-21

14-18 The Genova Water Hammer Muffler slips over any 1/2" water supply tube using a Genogrip Adapter. Genova All Purpose Seal Lubricant is used on the tube end to make the installation easier. Broad wings let the muffler be screwed sturdily to a header or to a wall.

14-20 A muffler can be added to any fixture or appliance to stop water hammer. Do it simply by cutting through the water supply stubout, installing a tee in it, and sliding the muffler over a stub coming from the tee's branch.

14-21 A replacement kitchen sink installation, with floor water supply and Part No. 540051 Universal Line Valve fitted to the cutoff copper tube, uses a Water Hammer Muffler to prevent pipe damage. They're installed with a Universal Tee (Part No. 545051) below the fixture supply valve.

Water filter

The strong twisting force during removal and installation of water filter canisters to get at their cartridges inside can be murder on piping. When the piping is all that's holding a filter, it can be strained to the point of leaking or breaking. You could end up with a disaster. To prevent this, a filter unit should be mounted firmly to a wall or framing. That way the building, not your plumbing, takes the twisting force of canister removal. Filter mounting would be easy if manufacturers always furnished a means of doing it. Only a few do. Fig. 14-11 to 14-17 show proper filter mounting.

Stopping water hammer

Houses in which the plumbing pounds angrily whenever a faucet or valve is closed quickly have a potentially destructive problem called

water hammer (see chapter 1). It can be cured with a Genova Water Hammer Muffler (Part No. 530901; see Fig. 14-18). Fig. 14-19 illustrates how it works.

A 1/2" push-on, hand-tighten Genogrip Adapter at the muffler's open end lets it be easily installed to the cold or hot water supply system, whether rigid or flexible tubing. Water test to be sure you have no leaks. Use the instructions that come with it.

Install the chambers as close as possible to the shutoff valve. By using the Genogrip Adapters, you can reactivate nonworking air chamber tubes by cutting the tubing off 2" above the tee and slipping Water Hammer Mufflers over them. Figs. 14-20 and 14-21 show ways to use Water Hammer Mufflers, and they work in any position from straight up to straight down. Because

of its unique design, the Genova Water Hammer Muffler cannot become waterlogged, and should seldom if ever need recharging. The muffler works best close to the fixture or appliance, but will do some good even when placed in hot and cold water mains or in a house water service entrance. The idea is to locate mufflers without tearing out walls, floors, or ceilings.

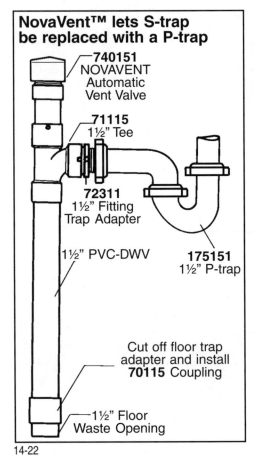

NovaVent™ lets S-trap be replaced with a P-trap

740151 NOVAVENT Automatic Vent Valve

71115 1½" Tee

72311 1½" Fitting Trap Adapter

1½" PVC-DWV

175151 1½" P-trap

Cut off floor trap adapter and install 70115 Coupling

1½" Floor Waste Opening

14-22

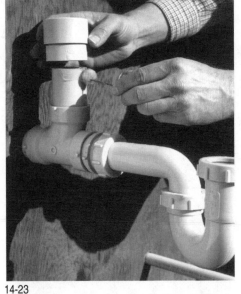

14-23

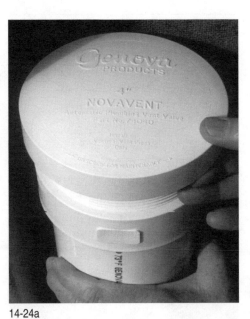

14-24a

14-23 Putting a tee in a fixture's waste line lets a NovaVent™ Automatic Plumbing Relief Vent Valve be used to prevent trap siphoning. It solvent welds over a 1½" PVC-DWV pipe. Make the pipe between the tee and valve at least 4½" long. This is to get the valve up high enough for it to work effectively.

14-24a, b The NovaVent™ Automatic Plumbing Relief Vent Valve admits air under vacuum conditions that would otherwise siphon a trap.

NovaVent™ automatic plumbing relief vent valve

Vent Stack

Lavatory

Bath

Toilet

P-Trap

14-24b

Curing trap siphoning

A fixture that loses its water seal by siphoning—caused by defective plumbing vents—can be cured without much effort. Installing a NovaVent™ Automatic Anti-Siphon Valve in the fixture's drain is the solution. It also promotes full drain flow by eliminating air locks. The one-way automatic air valve lets fresh air in, but closes when drain flow stops, to keep sewer gas in its place.

The vent is available as a 1½"

NovaVent™ Automatic Anti-Siphon Valve (Part No. 740151) and a 4" NovaVent™ Automatic Plumbing Relief Valve (Part No. 740401). The NovaVent™ anti-siphon valve is suited to washbasins, bathtubs and showers, sinks, laundry tubs, washing machine waste pipes and branch drains. The NovaVent™ relief valve is suited for use on 3" and 4" diameter plumbing vent stacks and can serve toilets as well as complete bathrooms. For a single-family residence, both valves

prevent trap siphoning by admitting air when a vacuum occurs in the DWV system. The valves also close under pressure to prevent gases in the DWV system from escaping into the house.

Both valves should be installed vertically, the 1½" NovaVent™ Anti-Siphon Valve above the pipe being vented and the 4" NovaVent™ Relief Valve, above the overflow level of the highest fixture served by it. For example, an anti-siphon valve may be

14-25

14-26

14-27

14-25 If you provide it with a trap adapter, a laundry tub may be fitted with a tubular P-trap. A short flanged tailpiece is held to the drain with a slip jamnut and the trap fits over the tailpiece.

14-26 The neatest, fastest way to make a hole in the floor for an automatic washer's standpipe is with a 2" hole saw. Chucked in a 3/8" or 1/2" electric drill, it cuts clean. Other sizes of hole saws are available, too.

14-27 A drain hose from the washer slips inside the 1½" PVC standpipe. It discharges washer wastes to the house drainage system. To keep sewer gases from escaping, a 1½" P-Trap (Part Nos. 78215, 78315, or 78415— the last two with cleanouts) is solvent welded into the waste run below the floor.

installed under the counter of a fixture with a trap that tends to be siphoned. For instance, a troublesome S-trap can be replaced by a P-trap and a NovaVent™ as shown in Fig. 14-22. With a wall waste opening, make the installation as shown in Fig. 14-23.

The NovaVent™ relief valve may be installed in an attic above an add-on bathroom, as in Fig. 14-24, to save having to cut a vent pipe opening through the roof. Both valves must remain accessible. Follow instructions that come with the valve.

Automatic washer hookups
If you live in an older house without provision for an automatic washing machine, you can still have one. Select a location convenient to a vent stack or building drain where you can pipe the waste water from the washer. Give second consideration to nearness to a point where you can tap hot and cold water supplies. A 20-ampere 120-volt dedicated branch circuit should be installed to serve the new washer location. It must also have ground fault circuit interrupter (GFCI) protection.

Laundry tub. Since an automatic washer's drain pipe is the least flexible part of the hookup, make that run first. You may wish to add a laundry tub at the same time you plumb in the washer. It must be connected to a vented drain. The hookup is the same as for a kitchen sink. Use a 1½" PVC waste pipe and a vent or NovaVent™. A trap hookup is shown in Fig. 14-25. A laundry tub permits the use of a suds saving automatic washer. In any case, the washer empties into the tub.

Standpipe drain. You can also drain a washer directly into what is called a standpipe — a 1½" PVC pipe reaching 36" up through a hole in the floor behind or beside the washer (Figs. 14-26 and 14-27).

A solvent welded P-trap is installed below the floor. The same rules that apply to all traps apply to the standpipe and laundry tub trap. If the code-accepted distance is too far to a vent (see table 5-A), the trap must be branch vented. Easier yet, use a NovaVent™ on it.

Connection drain. The washer's drain pipe can tie into a vent stack using a tee. Connect to a horizontal drain by tapping in from above via a 45° elbow and a wye inserted into the horizontal pipe. Fig. 14-28 shows various possibilities for automatic washer waste hookups. It also depicts the components of CPVC water supply systems. Threaded iron hookups are shown on the hot water side. These are guides to the parts you will need for your own water supply system. A basement system would be most similar to the attic type hookup, which is supplied with water from above. (continued on page 109)

Two ways of adding an automatic washer

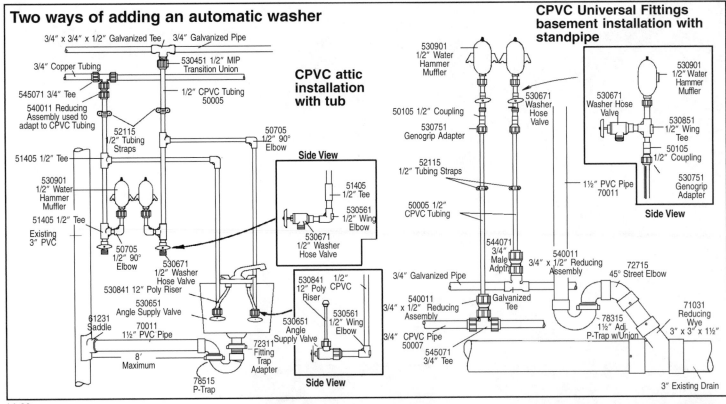

3/4″ x 3/4″ x 1/2″ Galvanized Tee   3/4″ Galvanized Pipe

3/4″ Copper Tubing

545071 3/4″ Tee

540011 Reducing Assembly used to adapt to CPVC Tubing

530451 1/2″ MIP Transition Union

1/2″ CPVC Tubing 50005

52115 1/2″ Tubing Straps

51405 1/2″ Tee

530901 1/2″ Water Hammer Muffler

51405 1/2″ Tee

Existing 3″ PVC

50705 1/2″ 90° Elbow

530671 1/2″ Washer Hose Valve

530841 12″ Poly Riser

530651 Angle Supply Valve

61231 Saddle

70011 1½″ PVC Pipe

8′ Maximum

78515 P-Trap

72311 Fitting Trap Adapter

50705 1/2″ 90° Elbow

CPVC attic installation with tub

Side View

51405 1/2″ Tee

530561 1/2″ Wing Elbow

530671 1/2″ Washer Hose Valve

530841 12″ Poly Riser

1/2″ CPVC

530561 1/2″ Wing Elbow

530651 Angle Supply Valve

Side View

CPVC Universal Fittings basement installation with standpipe

530901 1/2″ Water Hammer Muffler

530671 Washer Hose Valve

50105 1/2″ Coupling

530751 Genogrip Adapter

52115 1/2″ Tubing Straps

50005 1/2″ CPVC Tubing

544071 3/4″ Male Adptr

3/4″ Galvanized Pipe

540011 3/4″ x 1/2″ Reducing Assembly

3/4″ CPVC Pipe 50007

545071 3/4″ Tee

Galvanized Tee

540011 3/4″ x 1/2″ Reducing Assembly

1½″ PVC Pipe 70011

78315 1½″ Adj. P-Trap w/Union

72715 45° Street Elbow

71031 Reducing Wye 3″ x 3″ x 1½″

3″ Existing Drain

530901 1/2″ Water Hammer Muffler

530671 Washer Hose Valve

530851 1/2″ Wing Tee

50105 1/2″ Coupling

530751 Genogrip Adapter

Side View

14-28

14-29 A Genova Part No. 534991 Plastic Tubing Cutter can be used to nip off a sink or laundry tub's CPVC hot and cold water stubouts so you can tap into them for a nearby washer hookup. Leave at least ¾″ of tube protruding from the wall escutcheon for attaching the CPVC tee.

14-30 Solvent weld the ½″ CPVC tee in place replacing the fixture valve's stub in it. The tee can be used to begin plumbing a hot or cold branch run to the washer.

14-31 A pair of wing tees for the washer hose valves are mounted to a header between two studs. The wing tees make for a strong installation that will take the stress of valve handling. This hookup is supplied from an attic.

14-32a,b An absolute necessity when connecting an automatic washer is a pair of Water Hammer Mufflers— one in the hot and one in the cold water supply. These cushion the shock of a washer's fast-acting solenoid valve shutoffs. They'll work equally well in any position, even upside down as in 14-32a. Or they may be installed horizontally, as in 14-32b.

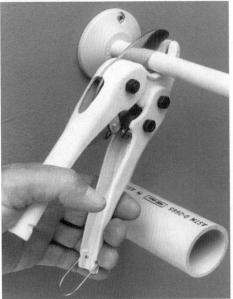

14-29

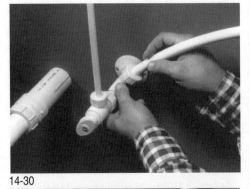

14-30

14-31

14-32a

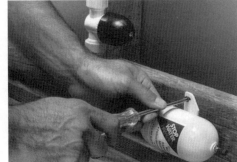

14-32b

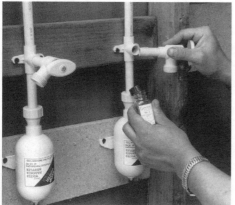

14-33

14-34

14-35

14-36

14-37

14-33 Washer Hose Valves get solvent welded to short stubs. Then they are, in turn, solvent welded into the wing tees. Placing them at an angle suits the natural hose alignment. If a finished wall were to be installed, this step would follow it. Then, Genova Torque Escutcheons would be used between wall and valves.

14-34 The washer hoses fasten right up to the threaded hose bibbs and reach to the washer's inlets. As always, the cold supply goes on the right, the hot on the left.

14-35 A food waste disposer replaces the right-hand drain basket of a double-bowl kitchen sink. Remove the old basket and clean the area, replacing it with the disposer's upper flange set in plumber's putty. Connect the rest of the mounting hardware from below. The last step is to clean up excess putty from around the upper flange.

14-36 The disposer then mounts beneath the sink. Because brands vary, follow the specific directions with your unit. End with the unit's waste opening facing the opposite sink bowl.

14-37 Drain attachment is the next step. This one uses a flanged tailpiece, surrounded by a spongy washer, and clamped. Other disposers use extension tailpieces. These contain their own slip couplings.

The left-hand drawing shows the use of Part No. 61231 Saddle Tee. If your vent stack is 3" Genova In-Wall Schedule 30 or 3" Schedule 40, this easy method may be employed. In any case, the fittings you use to make the connection depend on what kind of pipes you are connecting to.

Connecting water supply. For the washer's water supply, a pair of 1/2" CPVC hot and cold water pipes are run to the wall behind the washer. If the washer is next to a sink, its water supply can often be tapped as shown in Figs. 14-29 and 14-30. In building the system, make use wherever needed, of wing elbows and wing tees (Fig. 14-31). In conjunction with the wing tees you'll need two Water Hammer Mufflers.These prevent damage to the water supply system when fast-acting washer solenoid valves snap shut at the end of each fill or spray cycle.

Food waste disposer

A food waste disposer can be added easily to a double-bowl kitchen sink that was put in without one. Because Genova's continuous waste system is designed to accept a disposer, two waste outlets at the wall are not needed.

When installing a disposer, it's probably simplest to remove all the sink's drain hardware, and replumb with new. You don't want corroded, passage-restricted drain tubes holding back the flow of disposer wastes. Follow the instructions that come with the disposer. A dedicated switched 20-ampere 120-volt branch electrical circuit is needed. A disposer may be plugged into a switched receptacle or directly wired through a switch. In any case, proper wiring and grounding practices as spelled out in the National Electrical Code should be followed. This book does not cover electrical work. We recommend having an electrician handle that part of the disposer installation for you.

A disposer usually hangs from the right-hand sink bowl. Its waste opening connects to either a Flanged Tailpiece (Part No. 138121) or an Adjustable Tailpiece (Part No. 138301), which reaches over to enter the tee or wye of the continuous waste system. One piece of the continuous waste package will not be needed, as the disposer and tailpiece take its place. The major steps in plumbing a disposer are shown in Figs. 14-35 to 14-40.

14-38

14-39

14-40

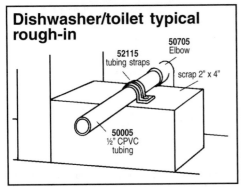

Dishwasher/toilet typical rough-in

50705 Elbow

52115 tubing straps

scrap 2" x 4"

50005 ½" CPVC tubing

14-41

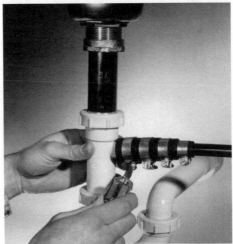

14-42

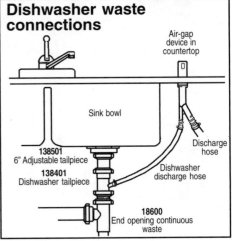

Dishwasher waste connections

Air-gap device in countertop

Sink bowl

138501 6" Adjustable tailpiece

138401 Dishwasher tailpiece

Discharge hose

Dishwasher discharge hose

18600 End opening continuous waste

14-43

14-38 Finish connecting the continuous waste system as normal with the disposer serving as the right-hand portion of it. This shows a center opening continuous waste setup. The upper dishwasher drain connection coming out from the disposer has a knockout shield for a dishwasher drain hose connection.

14-39 An end opening continuous waste setup is handled similarly to a center opening one, except that a longer tailpiece is needed to reach it.

14-40 If the kitchen sink has separate waste outlets for the bowls, the disposer is plumbed directly to the one on its side. The discharge elbow comes with the disposer and enters a P-trap.

14-42 A dishwasher tee lets the appliance's discharge hose empty into the kitchen sink drain. A handy multisized dishwasher drain adapter connects the hose to the tee when it isn't the usual 7/8" I.D.

Dishwasher

A little more complex to install is a built-in dishwasher. You can do it, following the instructions with the unit. It, too, calls for its own 20-ampere 120-volt electrical circuit.

The dishwasher's cabinet should be located where its drain hose can reach the kitchen sink drain. Either a box out at the rear of the dishwasher cabinet accommodates the appliance's utilities or they're routed beneath it and connected at the front.

Water Supply. A dishwasher gets hot water only. The supply can be piped from the nearest hot water main or it can be tapped from the kitchen sink's hot water stubout, as shown in Fig. 14-30. Many dishwasher instructions specify 3/8" I.D. or larger tubing for the water supply. You'll find that a 3/8" O.D. Poly Riser works beautifully.

First, locate the water outlet's rough-in to best reach the inlet valve. Fig.14-41 shows the typical low wall rough-in for a dishwasher (it works for most toilets, too). From the backflow-preventer, if required, use a Poly Riser long enough to carry water from the fixture supply valve to the dishwasher's solenoid valve inlet.

You can coil it, if need be, to reach. Join it there with a 3/8" MIP by 3/8" O.D. flare (or compression) adapter screwed into the appliance's inlet valve. Cut off the Poly Riser's shaped end and flare it. The flare nut will then join the riser tube to the adapter—watertight.

Drain. A specialized adjustable tailpiece is available for handling the wastes from a dishwasher. It's called a dishwasher tailpiece (part no. 138401), and it contains a slip coupling at the top end. Through the side is a 7/8" O.D. tee opening that connects to a 7/8" I.D. dishwasher discharge line. You install it in the kitchen sink's drain system. If your dishwasher uses a different sized discharge hose, a dishwasher drain adapter (widely available—Fig. 14-42) joins the appliance's drain hose to the dishwasher tee. And if the dishwasher is used in conjunction with a food waste disposer, you can plumb its drain into the knockout opening in the disposer. The opening is provided for that purpose.

Fig. 14-43 illustrates how a dishwasher's discharge hose is connected to a kitchen sink. Almost

Typical sump pump installation

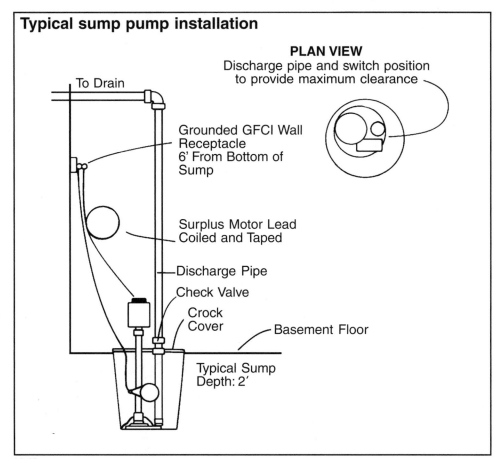

PLAN VIEW
Discharge pipe and switch position
to provide maximum clearance

To Drain

Grounded GFCI Wall
Receptacle
6' From Bottom of
Sump

Surplus Motor Lead
Coiled and Taped

Discharge Pipe

Check Valve

Crock
Cover

Basement Floor

Typical Sump
Depth: 2'

14-45

14-44

14-44 The best sump pump installation is in the floor. Mark around the sump crock and then chip out a hole in the floor large enough for it. This is mighty hard work, so if you can, rent a power jack hammer. Otherwise, use a cold chisel and hammer (but not a carpenter's hammer), and wear eye protection. Finally, dig out below the floor for the crock.

all plumbing codes require that an air gap device be used in the discharge run. The air gap mounts through a hole in the sink or in the countertop at the rear next to the backsplash.

Your dealer should be able to recommend one. If you don't need to use an air gap, at least route the discharge hose up to the lower face of the countertop and hang it from a wire loop there. From there it goes down to connect to the dishwasher tee. This up/down loop will prevent back-siphoning of sink wastes into the dishwasher. Test for leaks by running a load of dishes.

Sump pump

If you have a rural home with a basement, you'll probably want a sump pump to keep the basement dry. In adding a sump pump, you'll first need to create a pit for it by cutting a hole through the basement floor (see Fig. 14-44). Place a sump crock—a 2' length of 18" vitrified clay or concrete sewer tile laid spigot end down—in the hole. Or plastic sump crocks are available. These are much easier to work with. The sump pit should be large enough that the pump and its on/off switch don't touch the sides. If you make all

drains enter above the bottom, the bottom may be made of cast concrete. If seepage enters from below, the bottom of the sump pit should be gravel.

It's important that the pump be raised up off the bottom of the sump to keep collected sediment from jamming the pump impeller. Setting it on brick or the like will do the job.

Make a cover of 3/4" exterior-grade plywood. Holes can be cut in the cover for an upright sump pump, the discharge pipe, and power cord. If the cover is split into two halves, it can be easily installed around them.

Floor drains. Drains in the floor may be cut into the concrete, too. Make these of 1½" PVC-DWV pipe placed on a 1/4" per foot slope toward the sump pit. Use Part No. 78860 6"x 6" all vinyl bell traps in the floor at low spots. Where the amount of water to be handled by floor drains is likely to be considerable, as in a garage floor drain where cars will be washed, use a larger 9"x9" Part No. 78890 bell trap. This is made to fit both 3" Schedule 40 PVC-DWV pipe and 3" Schedule 30 pipe. And for really large amounts of water, as might be handled by an outdoor drain, use a Part No. 71450 PVC floor drain assembly. It takes a 4"

Schedule 40 PVC-DWV pipe or a 4" sewer and drain pipe. See chapter 17 for more information on these special products.

Lead the drains—footing drains, floor drains, and others—in through holes chiseled through the side of the crock. Make the sump pump's hookup as directed in the instructions with it. The pump's electrical connections need a 20-ampere three slot grounding type GFCI receptacle on a branch circuit that serves only the pump.

Pump hookup. Make the plumbing connections as shown in Fig. 14-45. If local codes require it, you'll have to install a fitting called a slop sink, which empties by gravity through a trap into a stack or drain pipe. If your code outlaws putting sump wastes into the sewer, you'll have to run them elsewhere, perhaps to a dry well (Page 132).

Sometimes, when only a basement laundry is to be drained, the sump crock is set on top of the floor.

Check Valve. Every sump pump installation should have a device called a check valve. This is to prevent wasteful backflow of water inside the discharge pipe after the pump shuts off at the end of each cycle. Genova Part No. 966311 Universal Check Valve

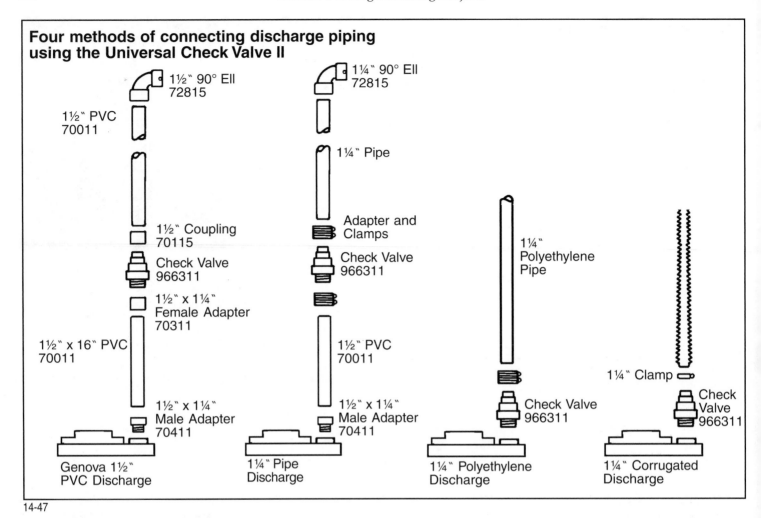

Four methods of connecting discharge piping using the Universal Check Valve II

1½" 90° Ell 72815

1¼" 90° Ell 72815

1½" PVC 70011

1¼" Pipe

1½" Coupling 70115

Adapter and Clamps

Check Valve 966311

Check Valve 966311

1½" x 1¼" Female Adapter 70311

1¼" Polyethylene Pipe

1½" x 16" PVC 70011

1½" PVC 70011

1½" x 1¼" Male Adapter 70411

1½" x 1¼" Male Adapter 70411

1¼" Clamp

Check Valve 966311

Check Valve 966311

Genova 1½" PVC Discharge

1¼" Pipe Discharge

1¼" Polyethylene Discharge

1¼" Corrugated Discharge

14-47

14-46

14-46 A Genova Universal Check Valve II is designed to fit just about any sump pump discharge setup and remain corrosion free in the awful atmosphere it lives in. It may be used with its flexible band-type couplings or installed directly to the pump. It's truly a universal check valve.

II (Fig. 14-46) is the finest available. The Check Valve II also prevents back-siphoned water from a flooded storm drain or culvert from flooding into your basement.

Fig. 14-47 illustrates how the Universal Check Valve II can be used in four different sump pump discharge hookups. It will solvent weld directly to a 1½" PVC coupling (Part No. 70115) and then you can continue from there with 1½" PVC-DWV pipe.

The valve also comes furnished with two flexible adapters and the necessary clamps for a flexible hookup. Fig. 14-47 shows the check valve adapting to 1¼" PVC pipe on the bottom. It then uses a Male Adapter (Part No. 70411) to reduce down to connect into the sump pump. The third illustration shows the check valve directly entering the sump pump, with 1¼" polyethylene pipe extending from the top of the check valve.

The Universal Check Valve II also fits 1¼" corrugated sump pump discharge hoses. A flexible hookup allows easy pump removal for servicing. It also tends to reduce sound transmission while the pump is running. A rigid discharge installation is superior, however, because it doesn't have flexible piping that could interfere with switch operation, possibly letting the basement flood.

Be sure that the pump is secured in the sump. See that it doesn't "walk" each time it cycles. And don't work with the pump-- or even handle it-- while it is plugged in.

Chapter 15 Adding a Half-Bath

Complete Half-Bath Plumbing System

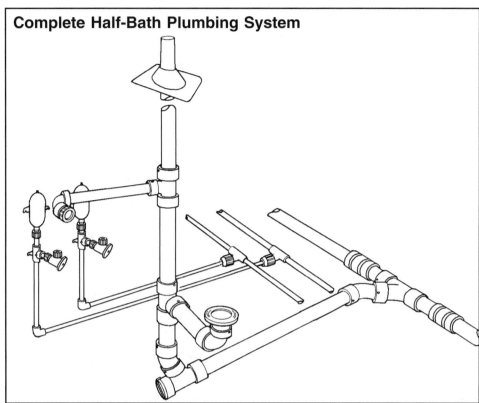

15-1

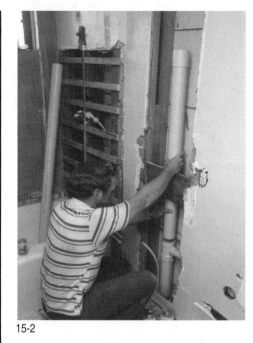

15-2

15-2 Plumbing for a half-bath usually calls for tearing out the walls to get access for vent and drain runs. Locate your vent run between studs as near to the toilet location as practical. Provide a reducing tee for the washbasin's waste line, then solvent weld the stack up and out through the roof.

15-3 Rough in the washbasin's waste pipe centered on it and 20″ above the finished floor. Cut access holes for the water supply tubes, as well.

15-4 Cut a 5½″ diameter hole in the floor for the toilet's soil pipe. It should center 12″ (or a little farther) out from what will be the finished wall.

15-3

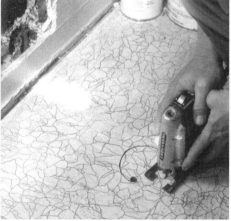

15-4

Adding a half-bath to your home will increase its value, as well as make it more livable. A half-bath consists of a toilet and a washbasin, which you can add at the end of a hallway, in a closet, or in the corner of a large bedroom. A space as small as 2½′ by 4′ will do, but that's really tight. It's better to have more room; the closer it is to existing plumbing, the better.

Planning

Before you begin hacking up walls and floors, sketch out a floor plan of your half-bath's fixture arrangement.

Plan for the toilet first; then plan the washbasin to fit. Tiny wasbasins as small as a foot square can be found, while the more normal sized ones stick out from the wall at least 1½′— washbasins can be wider.

Add a door and window(s) to your half-bath layout. If no window location is available, you can meet codes by installing a vent fan. This may vent through the bath's wall or ceiling. Final discharge may be through the house wall or roof. Ducting and a discharge vent are available where you get the fan unit.

If the bath is on the top floor, a skylight can be used to bring light into it.

Drains and vents

The new bathroom fixtures need draining and venting. They also need a hot and cold water supply. Whether you connect your new fixtures to a main vent that's already there or run a new vent up through the roof depends on how far away the present one is. Your local plumbing code will have something to say about the maximum distance. The

15-5

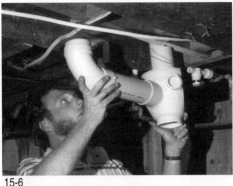

15-6

15-7

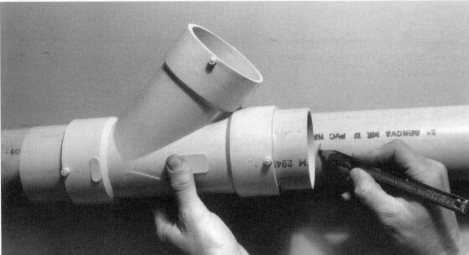

15-8

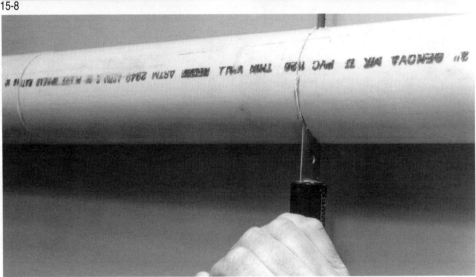

15-9

15-5 If the toilet's soil pipe will interfere with a joist, part of the joist can be cut out and boxed with headers. The nailed on headers should be of the same size lumber as the joists, and should support the remaining portions of the joist to the full joists on both sides of it.

15-6 It's a good idea to preassemble portions of plumbing, doing it where work is easy. Then each section can be installed as a unit where the working is hard. This is a Special Waste & Vent Fitting containing solvent welded toilet soil pipe, and part of the vent stack.

15-7 Work away from the toilet's plumbing in all directions, using it as the beginning point. Here its drain is being run toward the existing house drain.

15-8 To join a new DWV run into an existing one, or to make repairs, mark the portion of the old pipe to be cut out. This is best done by assembling the new section and marking directly from it.

15-9 Next, saw out a length of old pipe between the marks. This saw has a blade that reverses direction for sawing up to a wall. If the pipe mounts on a wall, you'll need one. Install a slip coupling as shown in Fig. 5-46.

distance is in **developed length**, which means the measurement along the centerline of the toilet's soil pipe and fittings. To keep this length of soil pipe from siphoning the toilet's trap, hold the slope to 1/4" per foot. The closer you put the toilet to its soil stack, however, the better. If you wish, a whole new soil stack can be installed to serve the new toilet.

Fig. 15-1 shows the DWV and water supply plumbing for a typical half-bath.

In many cases, installing a new vent up through the roof is the easiest method (Fig. 15-2). Or easier yet, you can install a NovaVent™ Automatic Plumbing Relief Vent Valve.

The toilet and washbasin wastes will drain by gravity into the existing building drain. Make the new to old drain connection with wye and slip couplings or No-Hub™ joints. The use of No-Hub™ joints is shown in Fig. 5-41. Remember that the flow must enter a drain in the same direction as the flow already in it. Entry from above is desirable, although entry from the side is permissible.

When plumbing for an add-on bath as when doing any plumbing avoid right-angle turns in drains except when using special long-sweep elbows. Be sure to also provide cleanouts for all horizontal runs of pipe that are not accessible for cleaning from the fixture location. Genova Twist-Lok™ Plugs are ideal for this.

If your code doesn't permit trap arm entries below a toilet's soil pipe entry, use a Special Waste & Vent Fitting. Its smaller entries go in above that of the

15-10

15-11

15-12

15-13

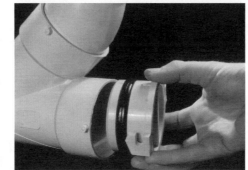
15-14

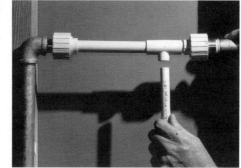
15-15

toilet. On the other hand, you can provide the washbasin with a drain and a branch vent. Easier yet, the use of a NovaVent™ (Part No. 740151) (see page 106) will vent the washbasin. It then can be drained below the toilet with no problem.

Water supply

Bringing the water supply into both fixtures with CPVC is the simplest part of the job. Start by bringing new hot and cold mains to your half-bath in 3/4" tubing.

Fig. 15-15 indicates a good way to tap into an existing threaded metal water supply run for your new half-bath. You'd need one in a cold main, another in a hot main.

To tap into a copper tubing water supply without doing any sweat soldering, you'd need a pair of Genogrip Adapters the same size as the copper tubing. Cut a short section from the copper main. Slip a Genogrip Adapter over each cut off end and install a CPVC tee by solvent welding. Take off from the tee with your new water supply run for the add-on bath.

Other chapters in this book, particularly Chapter 5, contain helpful information on plumbing a half-bath.

Sketch out both your DWV and water supply arrangements and order the necessary pipe, tubing, and fittings from your Genova dealer. He will be pleased to help you in choosing the right ones for the job. The wide selection of Genova Pipe and Fittings is on your side when it comes to adding a half-bathroom.

Saddle tee

For connecting a new 1½" or 2" drain to an existing 3" PVC-DWV or ABS-DWV plastic stack or drain, Genova markets one of the most useful fittings ever devised. It's called a saddle tee (Genova Part No. 61231). While the saddle tee is a Schedule 30 In-Wall fitting, it will work on 3" Schedule 40 PVC and ABS-DWV pipes as well. If you need it to fit a Schedule 40 pipe, be sure to wire it tightly in place while the solvent cement sets.

The saddle tee offers the simplest way imaginable to tap into an existing 3" plastic pipe. It incorporates the Genova Street-Socket Design. This accepts the branch pipe in either 1½" or 2" (see Figs. 15-16 and 15-17). A 1½" pipe goes right into the socket. A 2" pipe fitting—a coupling, usually—fits over the outside of the street/socket hub.

15-10 If you take measurements for pipe runs face to face of fittings, don't forget to add for makeup at each end of the pipe. PVC-DWV with its solvent welding joints installs so quickly that "hard time" working in a crawlspace is minimized.

15-11 Sometimes an old cleanout can be removed, allowing a straight-in run. Here a Part No. 60243, 4"x 3" Reducing Bushing, has been solvent welded over the old ABS drain pipe. It lets the new 3" Schedule 30 In-Wall drain be fitted.

15-12 When trying to make piping turns in tight quarters with street fittings, you can modify them by sawing off half the socket depth. The street elbow here also has had a similar amount cut from its end. The resulting bend will be tighter than when using the unaltered fittings. In making the reduced-depth joint, be sure to apply extra solvent cement. (Note: codes do not generally approve this trick, but it works.)

15-13 The vent stack should run out through the roof. Once the solvent cement has cured, you can saw the stack off 12" above the roof. It helps if someone holds the stack firmly while you saw.

15-14 Don't forget to install Twist-Lok™ Plugs where needed.

15-15 Here's how to tap water supply branch runs from existing threaded metal pipes. Remove a section of the pipe back to fittings on both sides. Then install MIP transition unions in the fittings. Cut a CPVC tube to fit between transitions and put a CPVC tee in it. The tee's branch then provides the spot for your new hot or cold water supply run.

15-16 Genova SaddleTee permits a 1½" or 2"
tap into a 3" plastic stack and drain. To use it,
give the back of the saddle tee and the pipe wall lots
of solvent cement.

15-17 Then place the saddle onto the pipe
right where you want the branch to enter. You can
hold it in place with mechanic's wire while the cement
cures. Finally, drill through the pipe wall inside the
tee's tapping and clean up with a half-round file.

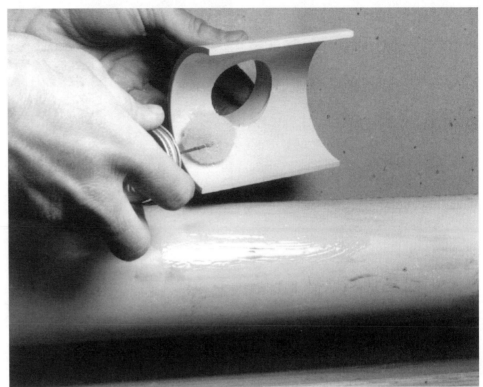

15-16

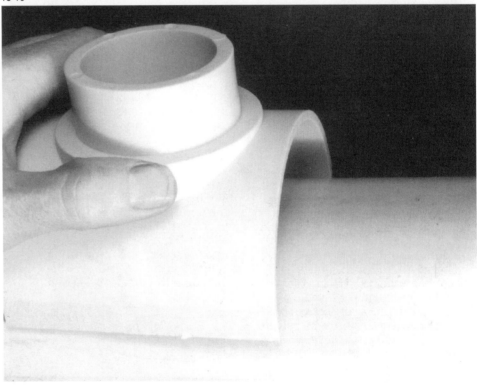

15-17

Chapter 16 Easy Lawn Sprinkler System

Master Plan of a Sprinkler System

-shows smaller (½″) pipe placement   =shows larger (¾″) pipe placement

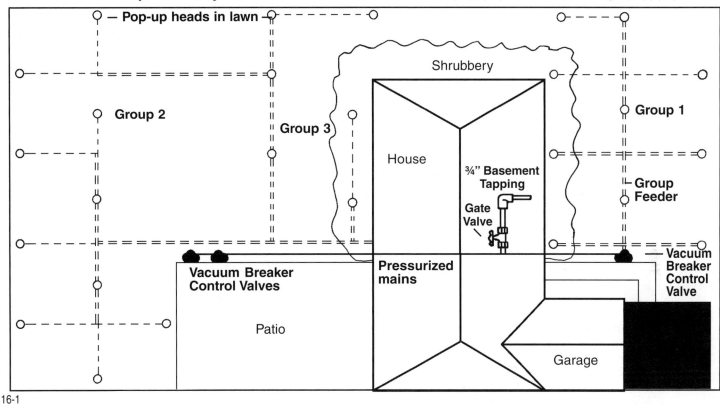

Pop-up heads in lawn

Group 2

Group 3

Shrubbery

House

¾″ Basement Tapping

Gate Valve

Group 1

Group Feeder

Vacuum Breaker Control Valves

Pressurized mains

Vacuum Breaker Control Valve

Patio

Garage

16-1

A great addition to your home plumbing is a lawn sprinkling system. Digging is the only tough part of the job. You can rent a ditch digger to help with this.Use easy solvent welding Genova 300 Series PVC pressure pipe — outdoor, cold-water only in plain or handy belled end — and fittings. Or you can use Genova 350/360/370/380 Series steel and plastic insert fittings for use with flexible polyethylene (outdoor, cold-water only) pipe. The rest is a cinch.

To offer what you need, the PVC pressure pipes and fittings and insert fittings come in sizes from 1/2″ to 2″. Additionally, Genova 300 Series PVC pressure pipes are available in pressure ratings from 125 psi to 850 psi (at 73°F) to suit any lawn sprinkler installation.

All 300 Series pipe sizes are designated according to nominal iron pipe sizing. For this reason, Genova 500 Series Hot/Cold CPVC systems, which follow copper tube sizing, cannot be accidentally interconnected. This helps prevent the use of cold water only plastic piping in the hot water supply system inside the house.

Master plan. Before you can plan your sprinkling system, you'll need to select the brand of sprinklers you'll be using. Your dealer will have the sprinkler manufacturer's instruction sheet containing the facts on flow rate and coverage. This is the information you'll need for planning your system, selecting sprinklers, spacing them, sizing pipes, and doing the installation. Also, call the local building department to find out whether you'll need a permit.

Sprinklers should be grouped according to how much water can be supplied. This is governed by the size of your house water service entrance and the water pressure. Measure the pressure with a hose-bibb gauge or call the water company. The more water, the larger each sprinkler group can be. Do not mix different types of sprinklers in a group. All should be alike. For example, impulse sprinklers and spray heads should be in separate groups.

Genova makes all the necessary pipes and fittings. A helpful brochure is available free on the use of 300

Series pipes and fittings to make a sprinkler system.

Size-larger design

A tip from Genova. A lawn sprinkler system uses so much water that its piping is often taxed to capacity. To ensure that each sprinkler gets all the water flow and pressure it needs, we recommend the "size-larger" design concept. In this, while the system's main stop-and-waste valve is no larger than your house main line, the main sprinkler line taking off from the valve is one pipe size larger. For example, if your house has the typical 3/4″ service entrance, you'd use a 3/4″ gate valve (with waste) and a 1″ pipe for the main pressurized lines leading away from it. Use Schedule 40 or Schedule 80 pipes for pressurized lines, and thinner-walled, lower-cost SDR 21 pipes beyond control valves. At control valves, you can switch back to 3/4″ pipe. Reducing fittings will do this for you. At the point where only a few sprinkler heads are served, you can further reduce to 1/2″ pipe. A 1/2″ pipe will carry up to 6 gallons a minute of water flow, so if

Table 16-A
The following charts supply information on minimum pipe sizing requirements. They are based on the system being at the same elevation as the water source.

60 PSI at source
minimum 40 PSI at sprinkler head

	Pipe diameter					
	½"	¾"	1"	1¼"	1½"	2"
100'	12	22	44	85	120	230
200'	8	15	28	55	80	150
300'	6	12	23	44	65	120
400'	5	10	19	37	55	100
500'	4	9	17	33	48	90

Pipe Length (left axis)
Fig. indicates gallons per minute

45 PSI at source
minimum 35 PSI at sprinkler head

	Pipe diameter					
	½"	¾"	1"	1¼"	1½"	2"
100'	8	15	30	55	80	150
200'	5	10	20	36	53	100
300'	4	8	16	29	43	81
400'	3½	7	13	24	36	68
500'	3	6	11½	21	32	61

Pipe Length (left axis)
Fig. indicates gallons per minute

35 PSI at source
minimum 30 PSI at sprinkler head

	Pipe diameter					
	½"	¾"	1"	1¼"	1½"	2"
100'	5½	10	20	37	55	105
200'	3½	7	13	25	37	70
300'	2¾	5½	10	20	30	56
400'	2¼	4½	9	17	25	45
500'	2	4	8	15	22	40

Pipe Length (left axis)
Fig. indicates gallons per minute

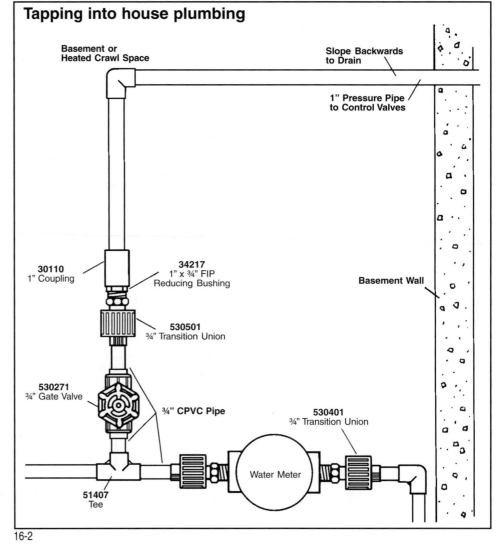

Tapping into house plumbing

Basement or Heated Crawl Space

Slope Backwards to Drain

1" Pressure Pipe to Control Valves

Basement Wall

30110 1" Coupling

34217 1" x ¾" FIP Reducing Bushing

530501 ¾" Transition Union

530271 ¾" Gate Valve

¾" CPVC Pipe

530401 ¾" Transition Union

Water Meter

51407 Tee

16-2

the remaining sprinkler-head capacities total less than that, they may be served by a 1/2" pipe.

Fig. 16-1 shows the principles of grouping of sprinkler systems. Fig. 16-2 shows size-larger design.

Installing the system

Where you tap from your house water supply system is important. Your tappings should come off the cold water main as far back toward the water meter as possible: a rule of thumb is anywhere upstream of a pressure regulator. This is to reduce sprinkler system interference with household water uses. In a freezing climate, be sure to slope the mains to drain back into a basement or heated crawlspace. Or you can use a gravel-filled pit with underground stop-and-waste valve and ground-level access box. Fit a threaded gate valve at the low point of the system. That way, the pressure lines can be drained for the off-season. Fig. 16-2 shows how

to handle a freezing climate tap-off under the house. In a frostline climate, main pressurized lines should be buried 6" to 12" below ground. Otherwise, 4" to 6" of depth is ample.

Vacuum-breakers. Each full flow control valve contains a vacuum-breaker—an anti-siphon device to prevent a cross connection between your house water supply and the sprinkler system. The vacuum breakers should not be under constant water pressure. To work properly, vacuum-breaker control valves—manual or automatic—should be at least 6" higher than the highest point of the line served. Connect to them with MIP Adapters (Part No. 30407, in typical 3/4"). Control valves may be together in groups which are then called control manifolds (Fig. 16-3).

Risers. From the control valves, run the nonpressurized group feeder lines. These pipes go 4" to 6" deep in V-shaped trenches through the center of each sprinkler's group

(Fig. 16-4). At selected spots, install tees up or laterals over to sprinkler heads (Fig. 16-5). Handy 1/2" x 6" Part No. 357605 polyethylene cutoff nipples rise from the group feeders or laterals. Tees, reducing tees, or elbows accommodate them. Risers may be cut to the desired height, enabling their length to be determined at the time of installation. Laying a piece of pipe across the trench on top of the ground will indicate the elevation of each, flush with the lawn sprinkler. Like pressurized mains, group feeders and laterals should slope to a drain spot.

After the last solvent welded joint has had ample time to cure, you can turn on the water and check the system for leaks. With a solvent welded system, leaks are very rare. If there are no leaks, the trenches can be filled in, the dirt tamped, and any sod replaced. You're finished.

Control Manifold

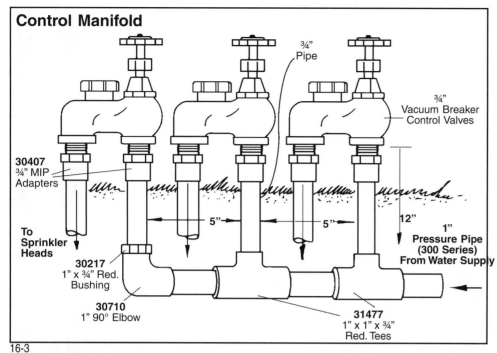

¾" Pipe

¾" Vacuum Breaker Control Valves

30407 ¾" MIP Adapters

To Sprinkler Heads

30217 1" x ¾" Red. Bushing

5"   5"   12"

1" Pressure Pipe (300 Series) From Water Supply

30710 1" 90° Elbow

31477 1" x 1" x ¾" Red. Tees

16-3

Pipes install in V-shaped trenches

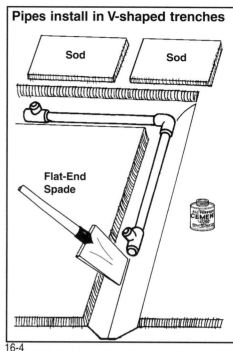

Sod   Sod

Flat-End Spade

16-4

16-5

16-6

16-7

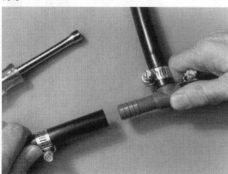

16-8

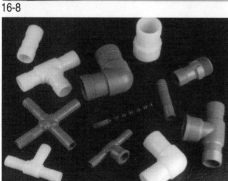

16-10

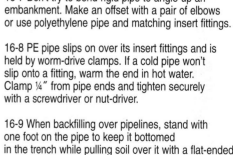

16-9

16-5 Piping runs install quickly using Genova 300 Series PVC Pressure pipe and fittings. Solvent weld a tee or reducing tee onto pipe branches.

16-6 Helpful Hint: use a pipe as a measuring guide (right). Wrap tape at proper spacing for sprinkler heads and cut other pipes at this point for installation of sprinkler tees.

16-7 Don't try to bend rigid pipe to angle up an embankment. Make an offset with a pair of elbows or use polyethylene pipe and matching insert fittings.

16-8 PE pipe slips on over its insert fittings and is held by worm-drive clamps. If a cold pipe won't slip onto a fitting, warm the end in hot water. Clamp ¼" from pipe ends and tighten securely with a screwdriver or nut-driver.

16-9 When backfilling over pipelines, stand with one foot on the pipe to keep it bottomed in the trench while pulling soil over it with a flat-ended shovel.

16-10 An assortment of types and sizes (½" - 2") of 350/360/370/380 Series insert fittings will suit any home sprinkler or underground water piping project. Fittings are offered in both polypropylene and nylon—use nylon for maximum strength.

"The Genova "Hub-Fit" Drain Grate

makes any Genova fitting into a drain with grating."

Chapter 17 Specialty Products in Vinyl

17-1

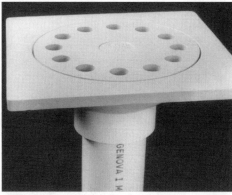

17-2

17-3

17-1 If you've ever tried to seal the joint between a plastic roof vent pipe and a metal roof flashing, you know despair. But slipping a Genova "Snap-Fit" Roof Flashing Cap down over the stack atop the old metal base flashing will end leaks.

17-2 A Genova 6"x 6" Bell Trap solvent welds to 1½" or 2" PVC-DWV pipe. A larger 9"x 9" version fits either a 3" Schedule 40 fitting socket or a 3" Schedule 30 pipe end.

17-3 Genova vinyl "Hub-Fit" floor strainer fits flush into a same-sized fitting hub. It makes a floor drain out of any 3" or 4" Schedule 40 fitting. Debris is prevented from entering the drain.

17-4 Genova's drain products may be cast right into a concrete floor, as this demonstration cross-section shows. Right to left: 3" "Hub-Fit" floor strainer; 6"x 6" bell trap; 9"x 9" bell trap; and 8" grate x 4" spigot PVC floor drain assembly.

17-4

Genova makes a number of specialty products in vinyl. A few, like the SW&V Fitting and the copper-to-PVC adapter, have already been covered in detail. Here are more of them.

Roof flashing. Genova's "Snap-Fit" Roof Flashing solves the old problem of vent stack leaks. The two-part flashing consists of a vinyl base and cap. The "Snap-Fit" Roof Flashing Cap also may be used alone over an old, leaking metal flashing, as shown in Fig. 17-1.

Bell traps. Genova offers bell traps in two sizes: 6"x 6" and 9"x 9". The amount of water to be handled governs which size you use. The 6"x 6" is shown in Fig. 17-2. Both install easily, making direct replacements for cast iron and die-cast bell traps.

"Hub-Fit" drain grate. The Genova "Hub-Fit" Drain Grate makes any fitting into a drain with grating (Fig. 17-3).

PVC floor drain. Genova's PVC Floor Drain is an 8" diameter high volume drain for use in floors and

garages. The PVC floor drain assembly is used where lots of surface water must be handled. Genova Part No. 71450 fits 4" PVC pipe. Fig. 17-4 illustrates four Genova drains.

PVC roof drain. Also fitting 4" PVC pipe, the PVC roof drain assembly has a raised grating to strain out debris. This part, No. 71460, is for use on flat roof buildings for roof drainage. It also may be used to make ground-level drains where lots of debris, such as leaves, is to be handled without clogging.

17-5 Backed up sewage is prevented by this back-water valve. It solvent welds directly to the building drain or house sewer without needing any adapters.

17-6 Genova Blue Bins are good at collecting items you need regularly and keeping them handy for use. They're also excellent organizers of things you might otherwise have to store in boxes.

17-7 Blue bins also come with integral dividers. These are especially handy for storing small parts.

17-5

17-6

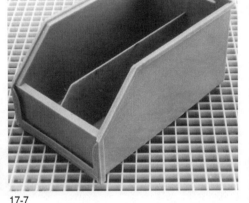

17-7

Shower Drain. If you build in a ceramic-tiled shower, an all vinyl Genova Shower Drain Fitting will save you lots of work and headaches. It has an adjustable height strainer and a lead pan retaining ring. Discharge size is standard shower 2" (Part No. 71470).

Other items
Here are brief descriptions of more special plumbing items you may need:

Toe Saver Floor Plug. A Toe Saver Floor Plug is a must for basement floor and slab-on-grade installations. Flush with the floor, it eliminates tripping over raised-head plugs. The Toe Saver comes in 3" or 4", Part No. 71853 or 71854.

Vertical Expansion Joint. If straight vertical DWV pipes are more than 35' long, some means for accommodating expansion and contraction may be needed. This fitting does the job. It's available in 2" to 4" sizes.

Backwater Valve. A one-way valve for sewage comes in 3" and 4", Part No. 77630 or 77640 (Fig. 17-5).

Blue Bins. Blue Bins are not a plumbing product, but are a useful product for the DIY plumber and home handyman. Genova's Blue Bins, home and shop organizing bins, will fit a ¼" peg board wall (Fig. 17-6). Each bin (Part No. 265325) provides more than 200 cubic inches of storage volume. Divided bins are also available (Part No. 265335).

Getting Genova Products
We didn't have space in this book to cover every product Genova makes for DIY plumbers. If you'd like to know more, just let Genova know and a current catalog of Genova products will be mailed to you. It shows the complete line of Genova products you'll need for plumbing your home.

Section 5 Doing Sewer-Septic Work
Chapter 18 Building a PVC House Sewer

18-1

18-2

18-3

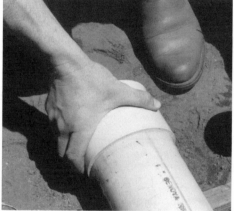

18-4

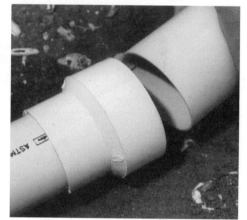

18-5

You can build your own PVC house sewer easily, provided you hire someone to dig and backfill the trenching for it.

The sewer begins 5' from the house foundation. It continues with solvent welded pipes to the septic tank or public sewer. If it's a sewer, a city or utility crew will make the final connection. You'll want to consult officials about the depth and location of hookup. Before calling for a hookup, also ask what fittings, if any, you are to provide.

What pipe?
You have a choice among pipes to make your sewer. The one we strongly recommend is 4" Genova Series 700 pipe (Fig. 18-1—Part No. 70041 in 10' lengths). If your code and number of house fixture units permit, you can save a little money by using the smaller 3" size (Part No. 70031). Both of these pipes are suited for PVC-DWV use, as well. If your sewer runs beneath a driveway or the water entrance piping runs in the same trench with it, the sewer

should be made of these thick-walled Schedule 40 pipes. Keep at least 12" vertical distance between the sewer and water lines, installing the water line on a solid shelf carved into the side of the sewer trench.

Genova 400 Series 4" vinyl sewer and drain pipes make a good sewer, too. Thinner-walled than Schedule 40, these pipes come in belled end, solid and perforated. Of course, you'll want to use the solid pipe (Part No. 40040) for a sewer. No couplings are needed. They meet or exceed the requirements of ASTM-D2729.

Fig. 18-2 shows both Schedule 40 and sewer and drain pipes side by side.

And for about the same cost as sewer and drain pipe, you can use 3" Genova Series 600 Schedule 30 In-Wall pipes for a sewer (check your code). These may also be used for DWV (Part No's. 60031/60032).

Building the sewer
Your sewer pipe should be sized according to the number of fixture

18-1 Tough 10' and 20' lengths of Genova 700 Series Schedule 40 PVC-DWV pipe join with solvent welding couplings (shown here).

18-2 Genova 400 Series vinyl sewer and drain pipe (left) is thinner-walled than Schedule 40 PVC-DWV pipe (right). The thinner-walled pipe is excellent for outdoor and underground use.

18-3 Solvent welding of pipe joints may be done either in the trench or on top of the ground. Out-of-trench welds need 12 hours of curing before lowering the pipes into their trench.

18-4 With both pipe end and fitting socket coated with Genova All Purpose Cement, join them with a slight twist and hold for 10 seconds. To keep the joint clean and dry, work with the pipe end supported off the ground on a block.

18-5 A 3" Schedule 30 In-Wall DWV system is joined to a 4" sewer made of 400 Series sewer and drain pipe with a Part No. 61543 4" x 3" Sewer Pipe Adapter.

Sewer Pipe Bedding

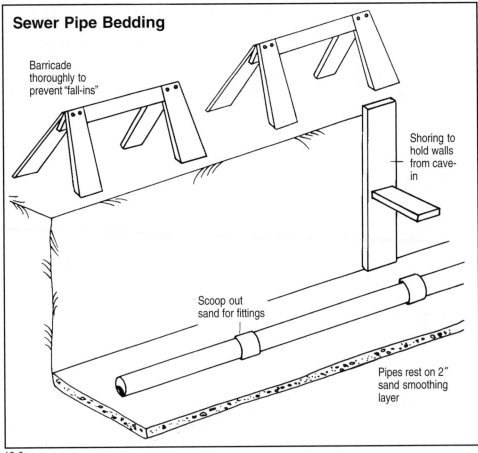

Barricade thoroughly to prevent "fall-ins"

Shoring to hold walls from cave-in

Scoop out sand for fittings

Pipes rest on 2" sand smoothing layer

18-6

units it handles. See Tables 5-B and 5-C for horizontal pipes. The ideal slope for a sewer is 1/4" per foot of run. In a pinch, the inspector might let you get by with a 1/8" per foot slope on a 4" pipe (but not on a 3" sewer). If the slope needs to be steeper to reach a low street sewer or septic tank, install the last 10' of pipe at the ideal slope.

The walls of a deep sewer trench must be shored to be certain they won't cave in. Hydraulically expanded shoring may be rented. Deep trenching may be necessary to reach a low-in-the-ground public sewer. Otherwise, deep trenches should be excavated with sloping walls.

Sewer pipes need proper bedding (Fig. 18-6). The easiest way is to use sand in the trench bottom. Avoid placing sewer pipes on earth backfill, however, as it compacts unevenly. This leaves low spots that later can cause clogging. Moreover, the sewer line should rest on its pipes, not on its fittings. Scoop out for couplings and other fittings. Also see that no pipe bridges across an unfilled depression in the trench bottom or crosses over an unexcavated bump.

Cleanouts. Most plumbing codes require a sewer cleanout. This may be located inside the house near the connection between the house DWV system and sewer or (Fig. 18-7) outside at the beginning of the sewer. An outside cleanout is extended up to grade with a pipe. Normally, an outside cleanout is two-way, made as shown in Fig. 18-7. Cleanouts are required at no greater than 100' intervals along the sewer. They also may be needed at certain sewer bends (check your code). Cleanouts are made from the same material as the pipe they serve. All may be closed off with Twist-Lok™ Plugs. If the plug is installed in a fitting without projecting pins, such as a coupling, hold it in with sheet metal screws driven between the plug and its fitting (see Fig. 4-18).

Backfill. Get your final inspection and then fill around the sewer pipe with sand, fine gravel, or selected rock-free earth. If you use earth, tap it lightly around the pipe, seeing that no rocks come in contact with the pipe. Once the pipe has been covered, the final backfill merely needs to be free of large rocks and foreign objects.

Sewer Cleanouts

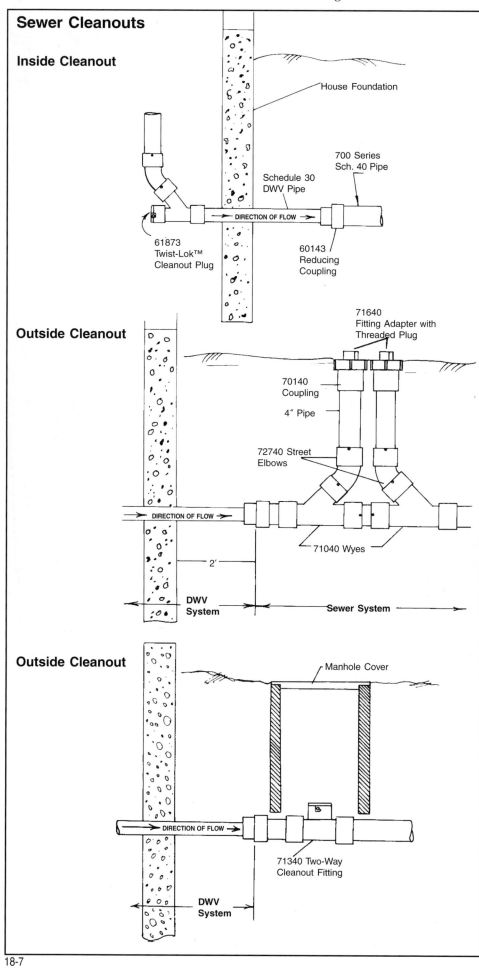

Inside Cleanout

House Foundation

700 Series
Sch. 40 Pipe

Schedule 30
DWV Pipe

DIRECTION OF FLOW →

61873
Twist-Lok™
Cleanout Plug

60143
Reducing
Coupling

Outside Cleanout

71640
Fitting Adapter with
Threaded Plug

70140
Coupling

4" Pipe

72740 Street
Elbows

71040 Wyes

← DIRECTION OF FLOW →

2'

DWV
System

Sewer System →

Outside Cleanout

Manhole Cover

→ DIRECTION OF FLOW →

71340 Two-Way
Cleanout Fitting

DWV
System

Table 18-A
Typical Sewer Clearances*

Horizontal

Distance from	Clearance
Buildings & Structures	2'
Property Lines	clear
Wells	50'
Streams	50'
Water Lines	1'
Public Water Mains	10'

*Check Local Code

Use the solid walled pipe for the sewer

*and distribution lines, the perforated pipe
for the seepage lines.*

Chapter 19 Building a PVC Septic System

19-1

19-2

Table 19-A
Septic Tank System
(Minimums)

House size	Tank Size
1 to 2 bedroom	750 gal.
3 bedroom	1000 gal.
4 bedroom	1200 gal.

For larger houses add 150 gallons to tank
for each additional bedroom

19-1 The purchased precast concrete septic tank
is lowered into a predug hole in the ground by the
truck that delivered it. Firms selling such tanks
can furnish details on the size of excavation needed.

19-2 Genova 400 Series pipes come two ways:
perforated for seepage lines (top) and solid-walled
for sewer and other tight lines. A complete
selection of fittings for both is also available.

If you live outside a city, you'll
probably need a private septic system
for sewage disposal. You stand to save
considerable money by building your
own system. Unless you enjoy digging,
you may want to hire someone to do
that part of the job. Before you start,
check with local public health
authorities for their requirements.

A disposal system consists of a sewer
line starting 5' from the house, a
septic tank, and a seepage system.
The previous chapter tells how to lay
the sewer portion.

The septic tank can be built in the
ground or purchased prebuilt and
lowered into a hole (Fig. 19-1). The
seepage system lets septic tank
effluent be absorbed into the ground.
The entire septic system goes below
ground, and piping slopes away from
the house for good gravity flow.

Genova's 400 Series Sewer Pipes
and Fittings are ideal for building a
sewer/septic system. Use the solid-
walled pipe for the sewer and
distribution lines, the perforated pipe
for seepage lines (Fig. 19-2).

Septic system design
Correct sizing of the septic tank is
important. The minimum tank sizes in
Table 19-A have enough extra
capacity to handle the wastes from a
garbage disposal, an automatic
washing machine, and other common
water-using appliances.

We recommend a two-compartment septic tank, as shown in Fig.
19-3. It functions more efficiently
than a single-compartment tank and
needs cleaning less often. If you
build the tank yourself, follow the

general details for the concrete block tank shown. Size it, figuring on about 7½ gallons per cubic foot.

A typical seepage bed is shown in Fig. 19-4. The seepage bed should be downsloped from the house for good gravity flow (check code). All elements of the septic system must be located to meet the minimum distance requirements shown in Table 19-B.

Making a percolation test
One important sizing question concerns the seepage field area. For proper effluent disposal, septic field size depends both on the amount of effluent to be disposed of and the ability of the soil to absorb. The number of bedrooms a house has is a good guide to the amount of effluent to be handled. But the best way to find your soil's absorptive ability is to test it. Tight clay soils absorb little; loose sandy soils absorb lots. Some health regulations call for such a test, called a percolation test, or simply "perc test." Here's how you can make your own.

Dig at least six test holes around the area you plan to use for the seepage field. Holes may be 4" to 12" in diameter. Dig to the same depth your seepage bed will be—often 36"—spacing holes uniformly over the whole seepage bed area. Roughen the sides of each hole to help absorb water. Remove loose dirt. Add 2" of sand or gravel to the bottom of each hole.

Now run at least a 12" depth of water into every hole. Add more water, as needed, to keep the level above the bottom for at least four hours.

Rate of fall. Finally, adjust the water level in each hole to 6" above the bottom. Then measure the drop in water level over a 30 minute period. Multiply that by two to get inches of fall per hour. That's your soil's percolation rate. Average the rate for all holes.

Good soils. If water will not stay in the holes for 4 hours, you may use this alternate measuring procedure: Add water to a 6" depth. Add more water as often as needed to keep it about 6" deep for three and one half hours. Then measure the drop in level during the next 30 minutes. Double that to arrive at the per hour perc rate.

Best soils. In really fast-perking sandy soils, you need to hold the levels for only 50 minutes at 6", measuring the level's drop over a period of ten minutes following. Multiply by six to get your perc rate. Fig. 19-5 translates these figures to

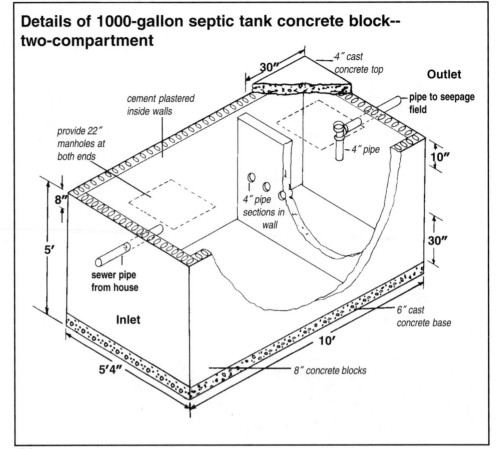

Details of 1000-gallon septic tank concrete block-- two-compartment

4" cast concrete top

30"

Outlet

pipe to seepage field

cement plastered inside walls

provide 22" manholes at both ends

4" pipe

10"

8"

5'

4" pipe sections in wall

30"

sewer pipe from house

Inlet

6" cast concrete base

5'4"

10'

8" concrete blocks

19-3

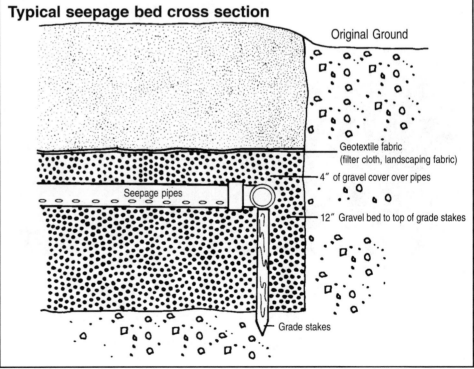

Typical seepage bed cross section

Original Ground

Geotextile fabric (filter cloth, landscaping fabric)

4" of gravel cover over pipes

Seepage pipes

12" Gravel bed to top of grade stakes

Grade stakes

19-4

Table 19-B
Septic System Clearances*

Horizontal Distance from	Sewer	Septic Tank	Seepage Field	Seepage Pit or Cesspool
Buildings & structures	2'	5'	8'	8'
Property lines	clear	5'	5'	8'
Wells	50'	50'	100'	150'
Streams	50'	50'	50'	100'
Trees	---	10'	10'	12'
Seepage Fields	---	5'	4'	5'
Water Lines	1'	5'	5'	5'
Distribution boxes	---	---	5'	5'
Public Water Mains	10'	10'	10'	10'

*Check local code

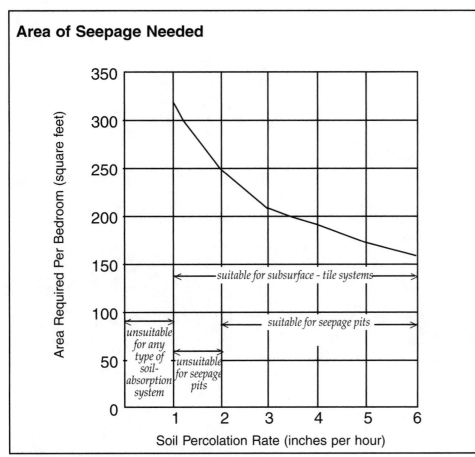

Area of Seepage Needed

suitable for subsurface - tile systems

suitable for seepage pits

unsuitable for any type of soil-absorption system

unsuitable for seepage pits

Soil Percolation Rate (inches per hour)

Area Required Per Bedroom (square feet)

19-5

seepage field size (check your code).

Worst soils. A tight soil with a perc rate of less than an inch per hour is poor for use as a seepage field. Find a better location. Or consult your public health officials on alternatives, if any. If water seeps into the holes as you dig them, the water table is too high for successful septic system operation.

An example: If the perc rate measures 3" an hour, you'll find from Fig. 19-5 that 200 square feet of absorption area is needed for each bedroom. A three bedroom house would thus need 600 square feet of seepage bed. That's an area 20' by 30'.

Septic tank

The septic tank decomposes wastes. It's the first part of the system to be installed.

Tank connections. The top of the septic tank should be at least 1' below ground, 2' to 3' in a heavy frost climate. Don't get the tank too deep, though, because its access holes must be dug up for inspection and possible cleaning every two years or so.

If the septic tank you get isn't equipped with pipe fittings at its inlet end, go right into it with the PVC sewer pipe. Mortar can be used to seal the space between pipe and a concrete or masonry tank. We recommend that no elbow or tee be used on the tank inlet. End fittings normally installed there invite plugging. At the outlet end of a fittingless septic tank, install a Genova Part No. 41140 Tee with a drop pipe, as shown in the Fig. 19-3. This should reach at least 18" below the liquid level. It can extend to within 1' of the tank bottom. The top of the tee is left open to vent the seepage field.

If your septic tank comes with inlet and outlet fittings already installed, you can adapt your pipes to them. Most often, clay pipe fittings are used. In this case, you connect to them with a Part No. 41540 Clay-to-Vinyl Adapter. To adapt to cast iron fittings, use a Part No. 41740 Vinyl-to-Cast Iron Spigot Adapter. Adapters are caulked to their fittings.

Seepage field

With health department approval, you may install seepage trenches, a seepage bed, or seepage pits. Trenches work a bit better than a bed because the sides as well as the (continued on page 131)

19-7

19-6

19-6 Seepage line construction starts with the hardest part--digging. Seepage trenches, shown here, require far less digging than a seepage bed. If you're working by hand, you'll surely want a trenched system.

19-7 Install grade stakes in the bottom of the 3′ deep trench. For good flow, the tops of the stakes should indicate a slope away from the septic tank. Same with the trench bottom. The stakes stick up about a foot.

19-8 A carpenter's level taped to a length of straight 2″ x 4″ long enough to span between grade stakes helps establish the slope. A 1/16″ shim (or drill bit) under one end of a 2′ level will give about a 3″ per 100′ slope with the bubble centered.

19-9 Put stones in the trench to the tops of the grade stakes. Stakes are used to help get a uniform slope. Stones should be ½″ to 2½″ in diameter.

19-10 A trenched seepage field needs a distribution box to divide effluent equally among two or more lines of tile. A precast concrete distribution box is leveled in the ground. The highest hole is the inlet; the lower ones are outlets.

19-8

19-9

19-10

19-11

19-12

19-13

19-14

bottom of the trenches absorb effluent. Seepage pits are deep holes in the ground lined on the exterior with porous masonry or concrete. They're used to solve special disposal problems, such as reaching down through a nonabsorbent surface soil to a subsoil where successful seepage can take place.

Trench-type system. Figs. 19-6 to 19-14 show how to put in a trench type seepage system. Join the septic tank's outlet tee through the wall of the septic tank with a 6' length of PVC solid-wall sewer and drain pipe. Use solid-wall pipe between the septic tank and distribution box.

To prevent effluent back up in seepage lines, they need to be sloped from 2" to 4" per 100'. That isn't much. (Some codes require that seepage lines be level.) No cleanouts are needed for seepage lines. No vents either, since house roof venting of the DWV system does that job.

Seepage trenches should be from 12" to 36" wide. Space them at least 6' apart, although 10' would be better. They're normally 3' deep and generally, no trench should be longer than 100'. Use several shorter trenches, instead. The seepage area of a trench is calculated as, width of trench times length, both in feet. On sloping land, seepage trenches should follow the contour of the ground. The smallest size stones that should be used in the trenches are 1/2", and the largest should be 2½".

Seepage bed system. Instead of

trenches, you can use a seepage bed. When the area for the seepage bed has been excavated, put in about 12" of gravel. Grade it out evenly. Build the seepage grid on top of the gravel. Install a Part No. 41440 Bull Nose Tee to the septic tank outlet. Set it level to start a grid seepage bed system, as shown in Fig. 19-15. The grid should be large enough to cover your entire seepage bed. Gravel fill and pipe-laying are the same as for seepage trenches, except that you're working on a large bed of gravel.

Backfill. You may need to have the system inspected. Then you can back-fill around the pipes with more gravel. Fill to at least 4" above them. Next, spread a complete covering of geotextile fabric (filter cloth), land-scaping fabric or untreated building paper to prevent soil infiltration. Finally, backfill with earth.

If properly designed, built, and maintained, your septic system should be effective for years.

If you have a good gutter and downspout system, such as a Raingo® one, you may want to build a storm water drain. This handles roof runoff below the ground. The system begins at the ends of downspouts. There Part No. 45234 adapts Raingo® Downspouts to Series 400 Sewer and Drain pipes or to Series 700 4" PVC-DWV fittings. Rarely does an outdoor runoff system come under plumbing codes, so the choice of pipe is yours. We suggest using thinner-walled Series 400 sewer and drain piping for this purpose. Either piping system can

19-11 Perforated seepage lines begin 5' from the distribution box. Lay them on the gravel touching the grade stakes with the perforations positioned at the bottom. When using Genova perforated pipe, the printing on the topside shows you that the perforations are down. Dry-coupling is permissible, but it's best to solvent weld the joints. This holds them together during backfill.

19-12 Fill at least 4" over the seepage pipes with more stones. Construction of a seepage bed is similar, except it uses a large hole and a gridwork of pipes.

19-13 A large rock or pipe cap placed at the end of a seepage pipe completes a run. This keeps stone backfill from entering the pipe. Such a closure is not necessary in a seepage bed gridwork.

19-14 Roll out a layer of building paper or geotextile fabric on top of the stones to keep subsequent soil backfill from infiltrating. Soil backfill may be mounded so that later settlement brings it level with the ground.

be laid below ground to carry off rainwater.

Lead the water to a good drain site. This may be a drainage ditch, collection pond, storm sewer, or dry well. Carrying the water away in piping prevents erosion. Rainwater should never be piped into a septic system, and rarely is it allowed in a public sewer. Also, avoid running it onto a neighbor's property. If necessary, build a dry well.

Underground runoff pipes should slope 1/4" per foot. Make the pipe line watertight by solvent welding the joints. Follow the pipe-bedding rules for sewer pipes given in the previous chapter.

A dry well is much like a septic system's seepage field: it depends on percolation for its operation. The dry well is a hole in the ground filled with gravel, crushed stones, or even scrap concrete blocks and bricks. You can use any material that has air spaces for the rainwater to filter through. The hole for a dry well should be at least one foot square and several feet deep. Keep it 10' to 20' from the house. Its top needs to be a minimum of 18" below ground and below the frostline, as does all piping. The bottom must be above the highest point of ground water. Keep the dry well 150' from a water well, 6' from a septic tank, and 20' from a seepage area, minimum. To avoid erosion, have a concrete splash apron where the concentrated flow enters your dry well. Fig. 19-16 shows a cross section of a typical dry well.

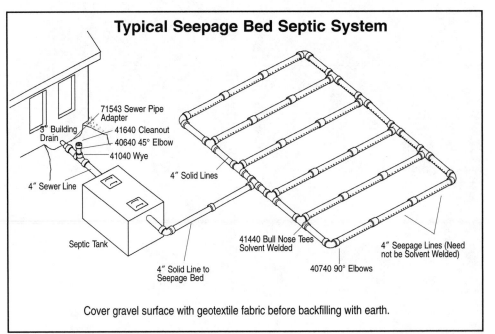

Typical Seepage Bed Septic System

71543 Sewer Pipe Adapter
3" Building Drain
41640 Cleanout
40640 45° Elbow
41040 Wye
4" Sewer Line
4" Solid Lines
Septic Tank
41440 Bull Nose Tees Solvent Welded
4" Seepage Lines (Need not be Solvent Welded)
4" Solid Line to Seepage Bed
40740 90° Elbows

Cover gravel surface with geotextile fabric before backfilling with earth.

19-15

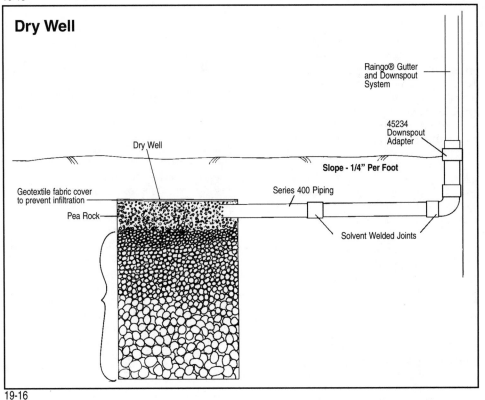

Dry Well

Raingo® Gutter and Downspout System
45234 Downspout Adapter
Dry Well
Slope - 1/4" Per Foot
Geotextile fabric cover to prevent infiltration
Series 400 Piping
Pea Rock
Solvent Welded Joints

19-16

Glossary

ABS—(Acrylonitrile-butadiene-styrene) A thermoplastic of low chemical resistance used in consumer items such as telephones, etc., as well as for building DWV systems. ABS has a low ignition point and it supports combustion.

Accessible—As applied to a fitting or valve, means being reachable for service, but which first may require the removal of an access panel or similar obstruction.

Adapter—A plumbing fitting that connects one kind of piping system to another kind.

Air gap—In a water supply system, the distance between the faucet outlet and the flood rim of the basin it discharges into. Air gaps are used to prevent contamination of the water supply by back-siphonage. "Air gap" is also used to describe a unique fitting installed to prevent back-siphoning in residential dishwasher drain lines.

All Purpose Solvent Cement—A solvent welding cement made by Genova that is designed to work on PVC, CPVC, ABS, and styrene pipe and fittings. Also called "universal" cement.

Angle-Stop Valve—A shutoff valve through which water makes a 90° turn. (See "fixture supply valve.")

ANSI—American National Standards Institute.

Anti-Siphon—Term applied to valves or other devices that eliminate back-siphonage.

Approved—Okayed by the plumbing official or other authority having jurisdiction.

Arcticweld® P—Genova solvent cement for use in freezing temperatures with PVC, ABS, and styrene.

ASME—American Society of Mechanical Engineers.

ASTM—American Society for Testing and Materials. U.S. governmental agency universally recognized for testing methods and performance criteria of materials, including plumbing pipe, fittings, and solvent cements.

Automatic Vent Valve—Genova product called "NovaVent™" that lets air into DWV system to prevent loss of trap seal. Takes the place of secondary vent stacks through the roof.

Backflow—Reverse flow of water or other liquids. Back-siphonage is a type of backflow.

Backflow-preventer—A device or means to prevent backflow.

Back-siphonage—Backflow of contaminated water by negative pressure in the potable water system.

Back Vent—Fixture vent connecting to a vent stack or drain stack.

Backwater Valve—A one-way valve installed in the house drain or sewer that prevents flooding of low-level fixtures from sewer back-up.

Baffle Tee—Tubular tee beneath one bowl of a double-bowl sink used for connecting a food waste disposer's discharge into a kitchen sink drain, thus utilizing a single drain trap. Removable baffle, when in place, prevents discharge from backing up into the other sink bowl.

Ballcock—Toilet tank bowl inlet water supply valve, which is controlled by a float ball. Usually of the anti-siphon type.

Bell Trap—A shallow-water-seal trap used for floor drains, primarily. Does not offer a sanitary seal, however, the trap feature is useful to catch small objects that might otherwise be lost down a drain.

Blowoff Valve—(See "temperature and-pressure-relief valve.")

Branch—Any part of a piping system other than a riser, main, or stack.

Branch Vent—A pipe installed specifically to vent a fixture trap. Connects with the vent system above the fixtures served.

Building Drain—The lowest house piping that receives discharge from waste and soil stacks and other sanitary drainage pipe and carries it to the building sewer outside the house.

Building Sewer—Normally begins about 5' outside the foundation of the house. Carries house sewage underground to a municipal sewer or private septic system.

Building Trap—A trap installed in the building drain to prevent sewer gases from circulating inside the house DWV system. This is no longer used in new construction.

Bushing—A fitting used inside the socket of another fitting to reduce its size to a smaller size of piping.

Cap—Used to seal off the end of a pipe permanently or temporarily.

Cesspool—Lined or covered excavation in the ground that receives domestic wastes from the drainage system. Retains organic matter and solids; lets liquid seep into the ground through its porous bottom and sides. Cesspools are currently outlawed by most codes.

Check Valve—A one-way valve for water flow. Used in a sump pump discharge line to prevent backflow into the sump pit with resultant short-cycling of the sump pump. A check-valve does not prevent cross connections.

Chlorination—Application of chlorine to water or treated sewage to disinfect or accomplish other biological or chemical results.

Cistern—Small covered tank chiefly used for storing rainwater for domestic use other than drinking. This is usually placed underground.

Cleanout—Accessible opening in the drainage system used to provide access for removing obstructions.

Common Vent—A vent connecting at the junction of two fixture drains that serves as a vent for both.

Continuous Vent—A vent that is a continuation of the drain it is connected to.

Coupling—Fitting used to join two lengths of pipe.

Corking Tool—A blunt-ended chisel-like tool used by plumbers for packing rope oakum and pounding plumber's lead in while making a caulked joint.

CPVC—Chlorinated polyvinyl chloride-- a heat-resistant rigid thermoplastic used for hot and cold water pipe and fittings that are joined by solvent welding.

Cross—Sewer and drainage type fitting used to connect two branch lines that approach a main sewer pipe from opposite directions. Used primarily in seepage bed grid piping in conjunction with perforated pipe.

Cross Connection—Physical connection between a drinking water supply and a drainage pipe or other source of contamination.

Developed Length—The length of line of pipe measured along the centerline of the pipe and fittings.

Double Tee (wye, etc.)—Used to bring two branch pipes into a single pipe.

Drain—Any pipe that carries waste water or waterborne wastes in a building drainage system.

Drainage Fitting—Fitting with insides designed for smooth flow of wastes by gravity. Features obstruction-free passages.

Drainage Systems—All the piping that carries sewage, rainwater, or other liquid wastes to the point of disposal or sewer.

Drum Trap—Old-style bathtub or shower trap largely replaced today by the P-trap.

Dry Well—Underground excavation used for leaching of other than sewage into the ground.

DWV—Stands for "Drain-Waste-Vent." The system that collects plumbing fixture discharges, and carries wastes out of the house.

Elbow—Fitting that changes the angle of two lengths of pipe.

Faucet—Fixture valve that opens and closes the flow of water to the fixture.

Faucet Nut—Half-inch nut that joins a fixture supply tube to the faucet tailpiece.

Female Adapter—Fitting with inside pipe threads (FIP) that permits connecting to male pipe threads (MIP).

Filter—Device for removing sediment from water and improving taste and odor. Sometimes used in a potable water system.

Finish Plumbing—Plumbing done after the building walls have been closed in--primarily plumbing fixture installation.

FIP—Female Iron Pipe Threads (See "Female Adapter")

Fitting—Plumbing part used to join two or more lengths of pipe. Fittings change directions, change sizes, branch out, and adapt.

Fixture Branch—A drain serving one or more fixtures that discharges to another drain or to a stack.

Fixture Drain—The drain from the trap of a fixture to a junction with any other drain pipe.

Fixture Supply—Water supply pipe that connects a fixture to a branch water supply pipe or directly to a main water supply line.

Fixture Supply Tube—Flexible tube that leads from the fixture shutoff up to the faucet tailpiece. (See "Poly Riser.")

Fixture Supply Valve—Valve located on wall or floor beneath a fixture, usually that turns off the hot or cold water supply to that fixture for repairs.

Fixture Unit, Drainage (dfu)—A measure of the probable discharge into the drainage system by various plumbing fixtures. In general, in small systems, one dfu approximates one cubic foot of water a minute.

Flashing—(See "Roof Flashing.")

Flex-A-Fitting Adapter™—Flexible new or replacement DWV fitting by Genova made of soft PVC. Allows for misalignment. Clamps onto pipe, or may also be solvent welded.

Flexible Tubing—Any tubing that is flexible in nature and used in water supply situations.

Flood Level Rim—The edge of a fixture receptacle from which water would overflow.

Floor Strainer—Prevents debris from entering a drain opening on the floor and fits flush into a fitting hub.

Flush Valve—A device located in a toilet tank that provides the "flushing" function.

Genogrip®—Patented Genova CPVC adapter for connecting CPVC or copper tubing.

Grade—The forward slope of a line of pipe in comparison to being horizontal. It is usually expressed as fall in fractions of an inch per foot of pipe length or inches per 100 feet.

Hub Adapter—Adapts cast iron pipe hub to thermoplastic piping. The hub adapter can be joined to cast iron with Genova Plastic Lead or molten lead.

Hydrostatic Tests—Tests of DWV system in which all openings are closed and the system is filled to overflow with water. This is done to test for leaks.

Individual Vent—A pipe installed to vent a fixture drain. It connects the vent system above the fixture served or terminates outside the building into the open air.

In-Wall Pipe—Designation for Genova Schedule 30 3" PVC-DWV pipe and fittings which fit within a standard 2x4 stud wall.

J-Bend—The U-shaped portion of a trap.

Line Stop—In-line valve that controls the flow of water through a pipe.

Long-Sweep—Designation for drainage fitting with a more gradual change of direction than a standard fitting.

Low-Heel Inlet—An inlet in an elbow used to connect a smaller branch or vent pipe at a change in direction in a larger vertical or horizontal drain line.

Main—Principal pipe to which branches are connected.

Main Stack (or Soil Stack)—Principal vertical toilet vent to which branch vents may be connected.

Makeup—The distance taken up by the fitting in a pipe-fitting joint. Also, the distance a pipe slips or screws into the fitting hub.

Male Adapter—Contains male pipe threads (MIP) to fit other pipes into a threaded pipe fitting.

MIP—Male Iron Pipe Threads (See "Male Adapter.")

No-Hub™*—A name given to hubless cast iron pipe that is joined with gasketed sleeve-type connectors rather than using caulked joints.

*No-Hub is a trademark of the Cast Iron Pipe Institute

Nonpotable Water—Water that is not safe for drinking.

Novaclean®—All purpose cleaner/primer for use in two-step solvent welding of PVC and CPVC.

NovaVent™—Name of Genova's automatic vent and anti-siphon valves that take the place of vent stacks. They let air enter the DWV system to prevent trap-siphoning, but keep gases from escaping into the house. Are utilized in many applications, including island-sink venting.

Novaweld® C—Genova solvent cement recommended for CPVC. This may also be used on PVC, ABS, and styrene.

Novaweld® P—Genova solvent cement recommended for PVC. This may also be used on ABS and styrene.

NSF—National Sanitation Foundation. An independent third-party certifying agency that assures products meet nationally recognized standards.

Offset—A combination of fittings that makes two changes in direction, bringing one section of pipe out of line but into a line parallel with the other section.

P-Trap—Trap in the configuration of a reclining "P" that lets water flow out, yet prevents the entry of gases and vermin.

PE—(Polyethylene) A flexible thermo-plastic pipe used only for cold water; out-of-doors and below ground.

PEX—(cross-linked Polyethylene) A flexible heat resistant tubing used for hot and cold water supply that is joined using Universal Fittings.

Pitch—(See "grade.")

Plastic—(See "thermoplastic.")

Plastic Lead—A scientifically formulated caulking lead substitute made by Genova. Can be used in place of molten lead in caulking applications. Has other uses, too. Expands as it cures.

Plumbing—The art of installing piping and fixtures. Used in installation, maintenance, extension, and alteration of all piping, fixtures, appliances, and appurtenances in connection with a plumbing system.

Plumbing Appliance—Special class of plumbing fixtures such as water heaters, water softeners, filters, etc.

Plumbing Code—A set of rules that govern the installation of plumbing.

Plumbing Fixture—A receptacle or device, which demands a supply of water that is either temporary or permanently connected to the water supply system. It discharges used water or wastes directly or indirectly into the drainage system.

Plumbing Official—The officer or other designated authority charged with the administration and the enforcement of the plumbing code. Or a duly authorized representative.

Plumbing System—Includes potable water supply and distribution pipe; plumbing fixtures and traps; drain, waste, and vent pipe; and building drains, including their respective joints and connections, devices, receptacles, and appurtenances within the property lines of the premises. It also includes water-treating equipment, gas piping, and water heaters and their vents.

Poly Riser—A flexible Genova fixture supply tube. Poly Risers are a direct replacement for flexible copper risers used to supply faucets and toilet tanks.

"Pop-Top"—The unique "Pop-Top" feature is used in standard Genova toilet flanges and prevents debris from entering plumbing system before setting the toilet. It also eliminates the need for installing a test plug during water-testing.

Potable Water—Water free from impurities sufficient to cause disease or harmful physiological effects. Conforms to the requirements of the U.S. Public Health Service Drinking Water Standards or the regulations of the public health authority having jurisdiction.

PP—(Polypropylene) This semi-rigid thermoplastic with high chemical and heat resistance is ideally suited for use in tubular drainage goods. It cannot be solvent welded; therefore must be joined by slip jam nut couplings.

Pressure Pipe—A cold-water pipe used to carry pressurized water for outdoor and underground uses such as lawn-watering.

Pressure Regulator—A device used in water service entrance lines to reduce line pressure for household use.

PVC—(Polyvinyl chloride) Rigid, chemically resistant thermoplastic used for DWV systems and cold-water-only pressurized systems outdoors or in the ground. It is joined by solvent welding.

PVC-DWV—PVC pipes and fittings approved for Drain-Waste-Vent use.

Reducer—Used to increase or reduce pipe size.

Relief Valve Drain—A pipe line leading from a water heater temperature-and-pressure valve, which often ends 6" above a floor drain.

Revent—(See "branch vent.")

Riser—A vertical pipe in the water supply system leading from the main.

Riser Tube—A short, flexible tube that connects the fixture to the water supply system (see "Poly Riser").

Rough-in—The installation of parts of the plumbing system that must be done before installing of the fixtures. This includes drainage, water supply, and vent piping and installation of the necessary in-the-wall fixture supports. Usually done before the walls are closed in. The term is also applied to the dimensions used for roughing in.

S-Trap—Trap shaped in the configuration of a reclined "S". Not currently allowed in new construction. Replacement use only.

Saddle Tee—A uniquely useful Genova fitting that solvent welds directly onto the side of a PVC-DWV or sewer pipe to provide a simple way of connecting a branch into a drain line.

Slip Jamnut—Nut used to join slip-fittings in tubular drains.

Soil—This is in reference to a pipe and means that it carries the flow from or vents a toilet.

Strap Wrench—A tool used to tighten appearance-sensitive fittings without marring them. A heavy strap around the part grips it.

Street—A fitting end that is pipe-sized rather than fitting-socket sized, allowing it to join into a fitting socket without an intervening length of pipe.

Street/Socket Design (SSD)—A unique Genova concept that permits a fitting to accept a pipe into its inside diameter or the next-larger size fitting over its outside diameter.

Styrene—Sometimes called rubber styrene (RS) or styrene rubber (SR). It is a lower-grade thermoplastic sometimes used for sewer and drainage piping, and joins by solvent welding.

Subsoil drain—A drain that collects below-ground water, carrying it to a place of disposal.

Sump—A tank or pit for sewage or liquid waste located below the normal grade of the gravity system and which must be emptied by mechanical means.

Sump Pump—An automatic water pump powered by an electric motor for the removal of drainage, other than raw sewage from a sump, pit, or low point.

Tailpiece—The short pipe leading from a faucet or sink drain.

Tee—Joins a branch pipe into a run of pipe.

Temperature-and-Pressure Relief Valve (T&P valve)—Installed in the top or side of a hot water heater tank to relieve a buildup of dangerous temperatures or pressures inside the tank.

Test Tee—A DWV fitting that installs in a vertical drain pipe to provide a cleanout opening close to the floor and flush with the finished wall. Used especially in slab-on-ground DWV installations to avoid having to make a special cleanout run beneath the concrete floor.

Thermoplastic—Plastic that hardens on cooling and softens on heating.

Toilet—A bathroom fixture used to dispose of human wastes. Also called "water closet."

Torque Escutcheon—Genova's solvent weld flange designed to prevent twisting damage to CPVC water supply stubouts to which valves or threaded transition fittings are attached.

Transition Union—A special fitting designed to compensate for the difference in expansion between thermoplastic and metal pipe in hot/cold water applications.

Trap—A fitting or device used to provide a liquid seal to keep sewer gases out of the house without affecting the outflow of sewage or waste water traveling through it.

Trap Adapter—Sometimes called a Marvel coupling, a trap adapter on the drain pipe joins tubular traps via a slip jamnut.

Trap Arm—The portion of a fixture drain between the trap's overflow weir and the vent. It is sometimes called a wet-vent, hydraulic gradient, or drain-and-vent.

Trap Seal—The vertical distance between the crown weir (uppermost point inside the dip of trap) and the overflow weir.

Tubing—Pipe that is sized nominally according to copper tubing sizes.

Tubular Goods—The fixture drainage pipes and fittings between a fixture and its trap adapter. These are thin-walled and joined by slip jamnut couplings.

Twist-Lok™—A unique Genova clean-out plug that provides an easily removable closure at practically any plastic DWV fitting hub.

Two-Way Cleanout Fitting—A fitting installed in a horizontal drain line which provides cleanout access to the drain in either direction.

Union—A fitting that joins pipes with a method for future disassembly without cutting them apart.

Universal Fittings—Unique Genova hot and cold water fittings that work directly with any type of tubing. Fittings join mechanically and leak-free.

Vacuum—Any pressure less than that exerted by the atmosphere.

Vacuum-Breaker—Device used in water supply line to let air in and prevent back-siphonage. Some are designed to operate under line pressure, others are not.

Valve—Fitting for opening and closing the passage for water.

Vent Pipe—Part of the DWV plumbing system which is open to the atmosphere and prevents both pressure and vacuum from building up within the system.

Vent Stack—Vertical vent pipe installed to keep the drainage system at atmospheric pressure. It protects trap seals from siphonage and back pressure.

Vent System—A pipe or pipes connected to a vent stack.

Vinyl—Short for polyvinyl chloride (PVC) or chlorinated polyvinyl chloride (CPVC).

Washbasin—A bathroom fixture also known as a lavatory or vanity.

Waste—Liquid-borne waste free of fecal matter.

Waste Pipe—A pipe that carries drainage from any fixture other than a toilet.

Water Closet—Toilet.

Water Distribution Pipe—Pipe used to supply water under pressure to fixtures and appliances.

Water Hammer—Sound created when water in a supply system comes to an abrupt stop when a valve closes quickly.

Water Hammer Muffler—A shock-absorbing device, made by Genova, that installs in the hot and cold water supplies at fixtures to prevent water hammer.

Water Heater—Plumbing appliance used to heat water.

Water Main—Public water supply pipe that feeds individual residential or commercial water service lines.

Water Meter—A device used to measure the amount of water that flows through it.

Water Softener—A plumbing appliance used to reduce the hardness of water, usually through the ion-exchange process.

Water Supply System—A water service entry pipe; water distribution pipes; and the necessary connecting pipes and fittings, control valves, and appurtenances in or adjacent to the building or premises.

Well—A driven, bored, or dug hole in the ground from which water is pumped.

Wet Vent—A vent pipe into which a fixture discharges at its base.

Wing Elbow, Tee—Sometimes called a drop-ear elbow or tee. They are used in conjunction with water supply stubout pipes or valves to provide a firm attachment for the pipe or a valve.

Wye—Used to connect a branch pipe into a drain pipe and provides a 45° angled entrance for the branch.

Index

Genova Products, Inc.

Meet Genova

Back in the mid-sixties my hardware dealer introduced me to Genova's CPVC hot and cold water supply piping system. The strong, rigid pipe could be cut with an ordinary handsaw and joined to its fittings by easy solvent welding. I tried it and was hooked. Now, any other method seems unnecessarily hard.

Later, in 1968, I featured CPVC in my first home plumbing how-to book. In researching for that book, I became acquainted with Robert F. Williams, then president. In the late 1950's, Robert F. Williams got the idea of utilizing plastics for plumbing systems which could be installed by individuals possessing no plumbing skills, and, as a consequence, Genova was born.

Williams' background as a licensed master plumber as well as an expert in the field of plastics made him eminently qualified to dispel the mysteries of plumbing to the average homeowner. Williams' goal was to make home plumbing something that homeowners could do easily themselves.

Since that early beginning, complete lines of Genova pipe and fittings have been developed by the senior Williams and his son, R.M. Williams, current Genova chairman and also a licensed master plumber. Mr. Williams lived to see his goal reached before he died in 1992.

Vinyl plumbing is not only easy to install, but the pure virgin vinyl utilized in Genova products lasts longer and works better than other piping systems. The company has continued to diversify into other labor saving products such as vinyl rain gutters and vinyl fencing.

Genova has grown from the idea of a single entrepreneur into a dedicated group of people committed to serving the needs of do-it-yourselfers everywhere, and those dealers who serve them. The corporate motto is "The People Who Get It Done!" Every employee is committed to the pursuit of that goal.

Other Genova Products to Help You

The quality and utility Genova builds into its do-it-yourself plumbing products extends to its products in other fields. Noncorroding Genova Raingo® gutters and downspouts and Genova Vinyl Fencing are designed for do-it-yourself home improvement. Made of solid vinyl, the parts will not rust, rot, or corrode, and never need painting, although you may paint them any color using ordinary house paint.

Raingo® Gutters and Downspouts. Before 1980, Genova pioneered Raingo® gutters and downspouts for do-it-yourself installation. At that time, rainwater systems for home use were either galvanized steel or aluminum. Because of its many benefits over metal gutters and downspouts, Raingo® soon dominated the home rainwater-handling market, and is still the leader.

Raingo® systems offer all the fittings you need to get the job done right, fittings like Bracket Spacers that allow a gutter to be attached to a sloping fascia.

Genova strap hangers let Raingo® gutters be fitted to any roof, whether it has a fascia or not.

And Genova drip edging slides up underneath shingles to extend an improperly made roof edge so that rainwater drips into the gutter, rather than behind it.

Then there's the Raingo® swing-up splashblock that leads water away from the house foundation yet stores up out of the way during lawn mowing and other activities.

What's more, Genova Repla K rain gutter and DuraSpout downspout is a vinyl clone to standard aluminum downspout, allowing it to be used as a replacement for a corroded metal downspout.

Genova gutters and downspouts come in full ten-foot lengths in both white and brown. They snap together for a fast, easy installation.

Ask your Genova Products dealer for an informative brochure on Raingo® (Part No. 252174). Also a 10-minute how-to-install-Raingo® videotape (Part No. 299023) is available from Genova by calling 810-744-4500.

Genova Vinyl Fencing. Genova's newest nonplumbing product line is its vinyl Fencing. As with Genova plumbing products and Raingo® gutters and downspouts, Genova Vinyl Fencing is designed for easy do-it-yourself installation. In fact, a Genova Vinyl Fence can be erected in just a few hours.

Either of two methods of securing posts may be used: For a solid installation, the 4" x 4" vinyl posts can be set in the ground in the usual manner. Then for additional strength, wood 4x4s may be lowered inside the vinyl posts or the posts may be filled with concrete. A decorative Genova Fence can be erected without digging a single posthole simply by driving standard steel fence T-stakes or U-stakes into the ground at post locations. Unique stake post mount fittings then marry the T-stakes or U-stakes to the vinyl fence posts, two fittings per post.

Genova Vinyl Fence offers great flexibility of design. It may be built either as a picket fence or as a rail fence. Moreover, the rails may be installed over the outsides of posts or installed between posts. For between-post installation, Genova rail post connectors, or routed post and notched rails greatly simplify the job. It all goes together so quickly you won't believe it.

Your Genova Products Dealer can provide a brochure on Genova Vinyl Fencing (Part No. 252261), or you can e-mail genova@genovaproducts.com. Furthermore, you are invited to visit Genova's website on the Internet at www.genovaproducts.com. A 12-minute videotape (Part No. 299032) is available to show how to install a decorative Genova Vinyl Fence.

Do -It-Yourself Plumbing Made Easy

About the Author

The author, like Genova, has been a pioneer in promoting successful and easy do-it-yourself methods for home plumbing. He was first to call the homeowner's attention to CPVC tubes and fittings almost 20 years ago. His four previous books on home plumbing all have brought the latest information on water supply with vinyl. This is his fifth plumbing book, a cover-to-cover new treatment of his previous "Plumb-It-Yourself, It's Easy With Genova." This book brings you the whole new story on using Genova vinyls.

A do-it-yourself writer in the home and workshop field for more than 40 years, Day has written numerous articles and books on home repairs and improvements. He also is consulted by publishers of do-it-yourself books. During the course of all this, he built three houses from the ground up, doing every task from bricklaying and concrete masonry to electrical and plumbing work. More recently, he helped a friend plumb a two-story, three bedroom house using all Genova materials from water supply to sewer connection.

Author Day lives in California with his wife Lois. He is active in the National Association of Home and Workshop Writers, which he helped to create.

Rich Day's clear, informative writing style makes a superb introduction to the world of easy do-it-yourself plumbing.